# Lighting & Electricity

## Consultants

**Ron Hazelton**, chief consultant for **HOW TO FIX IT**, is the Home Improvement Editor for ABC-TV's *Good Morning America* and host of his own home improvement series, Ron Hazelton's *HouseCalls*. He has produced and hosted more than 200 episodes of *The House Doctor*, a home-improvement series airing on the *Home and Garden Television Network* (HGTV).

On television, and in real life, Ron is a coach who visits people in their own homes, helping them do things for themselves. He pioneered the concept of on-location, home-improvement television, making over 600 televised house calls, doing real-life projects with real people.

The son of a building contractor, Ron has always had a fascination with the home and how it works. He left a successful career as a marketing executive to learn woodworking, eventually becoming a Master Craftsman and cabinetmaker. In 1978, he founded Cow Hollow Woodworks in San Francisco, an antique restoration workshop that restored over 17,000 pieces of furniture during his tenure.

**Rex Cauldwell,** is a contractor, master electrician, master plumber and author. He gives seminars on Home Inspection and writes articles for many national magazines including Fine Homebuilding, Tools of the Trade, American How-to, and Journal of Light Construction. He has authored books in publication, such as Wiring a House and Safe Home Wiring Projects.

**Merle Henkenius,** has 14 years experience as a master plumber, contractor (with great electrical knowledge), author, consultant and photographer. He has used this combination to write and photograph numerous home improvement articles for such magazines as Today's Homeowner, Popular Mechanic and Better Homes and Garden. In his free time he teaches home improvement at the local community college.

## Other Publications:

**Do It Yourself**
Home Repair and Improvement
The Time-Life Complete Gardener
The Art of Woodworking

**Cooking**
Weight Watchers™ Smart Choice Recipe Collection
Great Taste/Low Fat
Williams-Sonoma Kitchen Library

**History**
The American Story
Voices of the Civil War
The American Indians
Lost Civilizations
Mysteries of the Unknown
Time Frame
The Civil War
Cultural Atlas

**Time-Life Kids**
Library of First Questions and Answers
A Child's First Library of Learning
I Love Math
Nature Company Discoveries
Understanding Science & Nature

**Science/Nature**
Voyage Through the Universe

**For information on and a full description of any of the Time-Life Books series listed above, please call 1-800-621-7026 or write:**

Reader Information
Time-Life Customer Service
P.O. Box C-32068
Richmond, Virginia 23261-2068

HOW TO FIX IT

# Lighting & Electricity

By The Editors of Time-Life Books, Alexandria, Virginia

With **TRADE SECRETS** From **Ron Hazelton**

# Contents

CHAPTER 1
## Electricity in Your Home



CHAPTER 2
## Lamp Repairs



CHAPTER 3
## Lighting Fixtures



CHAPTER 4
## Wall Switches

Chapter 5
## Wall Outlets


Chapter 6
## Extending Circuits


Chapter 7
## Doorbells & Chimes


Chapter 8
## Outdoor Lighting


## Index

# FIX IT: Electricity in Your Home

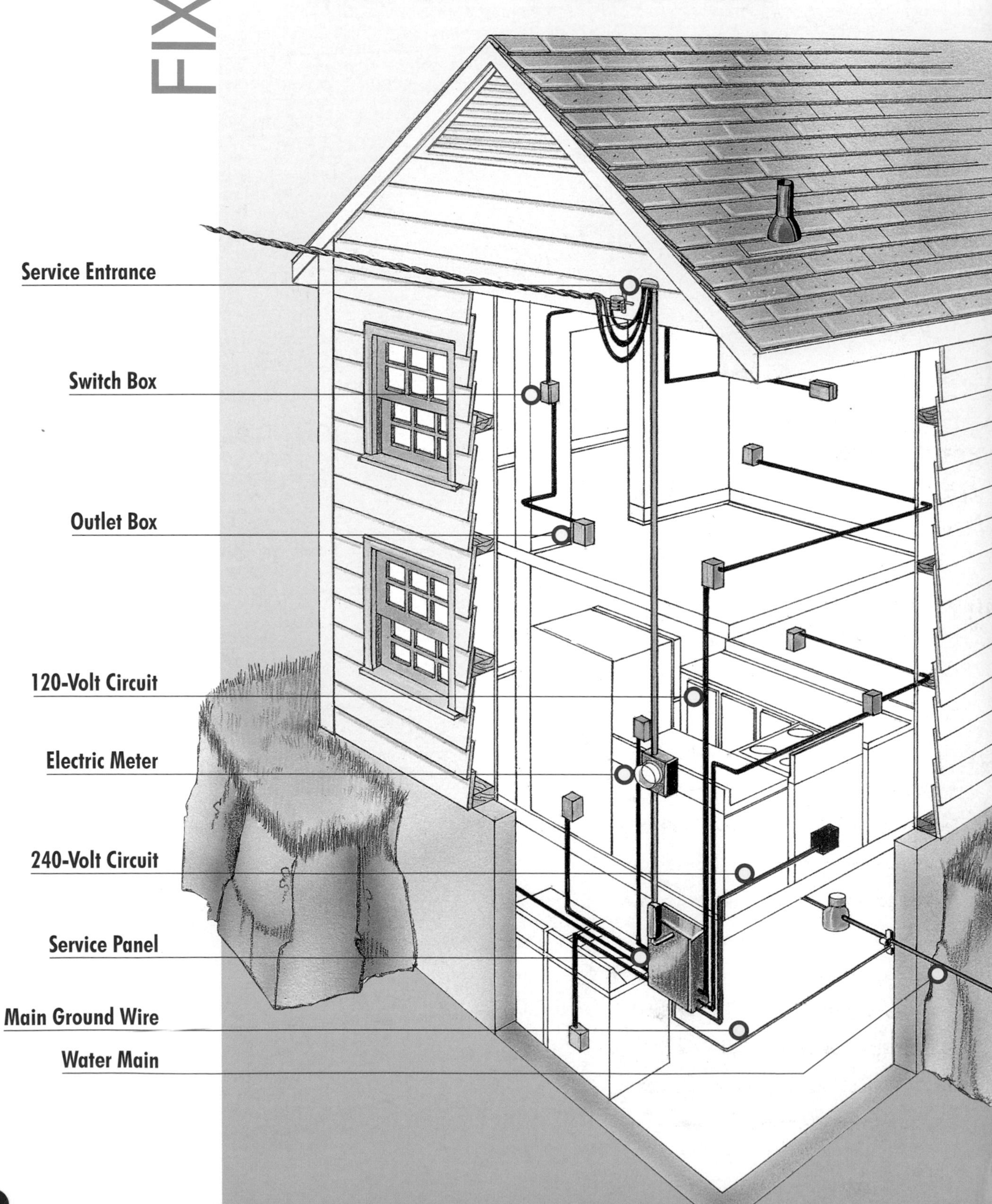

# Chapter 1

## How It Works

Electricity comes into a home through a service entrance. From there it passes through the electric meter to a service panel, which distributes power to individual circuits. Most homes have two kinds of circuits: 120-volt circuits carry enough current for small devices—lamps, TVs, blenders, and the like—while 240-volt circuits supply the additional power needed for large appliances like central air-conditioners, electric ranges, and dryers. Most circuits include a ground wire that links them to a water main or, in compliance with a more recent code revision, to two grounding rods buried in the earth at least 6 feet apart. All wiring connections must be made within a metal or plastic box.

There's a limit to the amount of electricity any given circuit can carry. For safety, circuit breakers trip or fuses blow when the demand exceeds capacity—or when there's a short circuit caused by a loose connection, a broken wire, or a malfunction within an appliance.

When problems occur, they're not difficult to diagnose and correct by studying the service panel *(page 10)*, evaluating the loads on the circuits *(page 13)*, inspecting the wiring connections, and testing individual components.

## Contents

# Troubleshooting

| Problem | Solution |
|---|---|
| • No power to the house | Check the service panel for a tripped main breaker or blown fuse **10** • Call your local utility company • |
| • No power to a particular circuit | Check the fuse or circuit breaker **10** • |
| • Breaker trips or fuse blows frequently | Identify all loads on circuit **13** • Compare capacity to total load **13** • Move one or more loads to another circuit • Check loads for a short circuit • Check wiring connections • |
| • Can't turn off power | Trace each circuit, and label breakers or fuses **12** • |
| • Arcs or sparks at the service panel | Don't touch the panel. Call the utility company to turn off power. Have an electrician check the panel before restoring power • |
| • Appliance, switch, or outlet sparks or is hot to the touch | Pull the power cord or kill power to that circuit **12** • |
| • Appliance is submerged or outlet is wet | Cut power to the circuit **10** • |
| • Basement or room flooded | Call the utility company to turn off power to the home • |
| • Power line fallen in yard | Call the utility company, police, or fire department • |
| • Someone stuck to live power source | Cut power to the circuit **10** • Use a wooden broom handle or chair to knock the person free **12** • |
| • Injury due to electric shock | Check whether the victim is breathing or has a pulse. If not, begin CPR if you are qualified. Call for help • |

# Before You Start

Understanding how power is distributed to your home will help you diagnose and repair most electrical problems.

### THE HEART OF THE SYSTEM

To begin understanding your home's electrical system, start at the service panel. If you don't know which breaker or fuse controls a given circuit in your home, take the time to map each circuit. Label each breaker or fuse on the panel with a number, and record them in a key to the circuit map. Identifying each circuit by its breaker or fuse, and making a record of every device on each circuit, can help you isolate and repair problems when they occur.

## Before You Start Tips:

- Fuses and circuit breakers prevent fires by keeping wiring from overheating. Never replace a breaker or fuse with another of higher amperage than the original.
- Use only lights, appliances, receptacles, and other electrical devices that have been approved by Underwriters Laboratory. The UL label indicates that a device has passed all standard safety tests.

Fuse puller
Flashlight

Fuses (if applicable)
Service-panel labels

Never stand on a wet or damp floor when working around electricity. When testing fixtures or removing fuses, keep one hand behind your back to prevent your body from making a complete circuit. Never replace a fuse or breaker with one of a higher amperage. Use tools with insulated handles.

# Controlling Power at the Service Panel

## Setting circuit breakers

A main service panel may have circuit breakers *(right)* or fuses *(below)*. The main breaker controls power to all circuits; individual breakers control separate circuits.

• Before attempting any electrical repair, find the breaker that controls the circuit you'll be working on *(page 12)*. Switch off power to the circuit by pushing the breaker toggle to the OFF position.

• If power to a circuit fails, check the service panel for a tripped breaker (a flashlight may be helpful). Some breakers have three positions—ON, TRIPPED, and OFF *(inset)*. Reset the breaker by first turning it off, then on. If it trips again, turn off one or two appliances and reset it.

• If the breaker continues to trip or the fuse blows repeatedly, call an electrician.

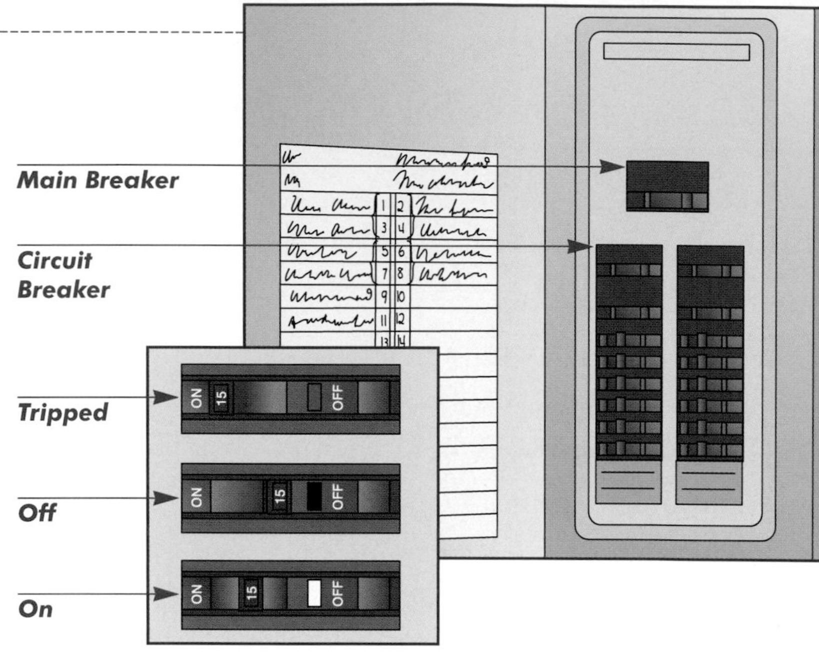


## Removing and replacing a fuse

• Before attempting any electrical repair, find the fuse that controls the circuit you'll be working on *(page 12)*. Cut power by removing the fuse. Unscrew plug fuses *(right)*. Deal with fuse blocks and the cartridge fuses that they hold as shown on the next page.

• If power to a circuit fails, check for a blown fuse (a flashlight might be helpful). A complete break in the metal strip inside a plug fuse indicates a circuit overload; discoloration indicates a short circuit *(inset)*. Cartridge fuses often show no sign of failure.

• To correct an overload, disconnect one or two appliances from the circuit and replace the fuse *(page 11)*. When you suspect a short circuit, find the problem and repair it before replacing the fuse.

• If problems recur, call an electrician.

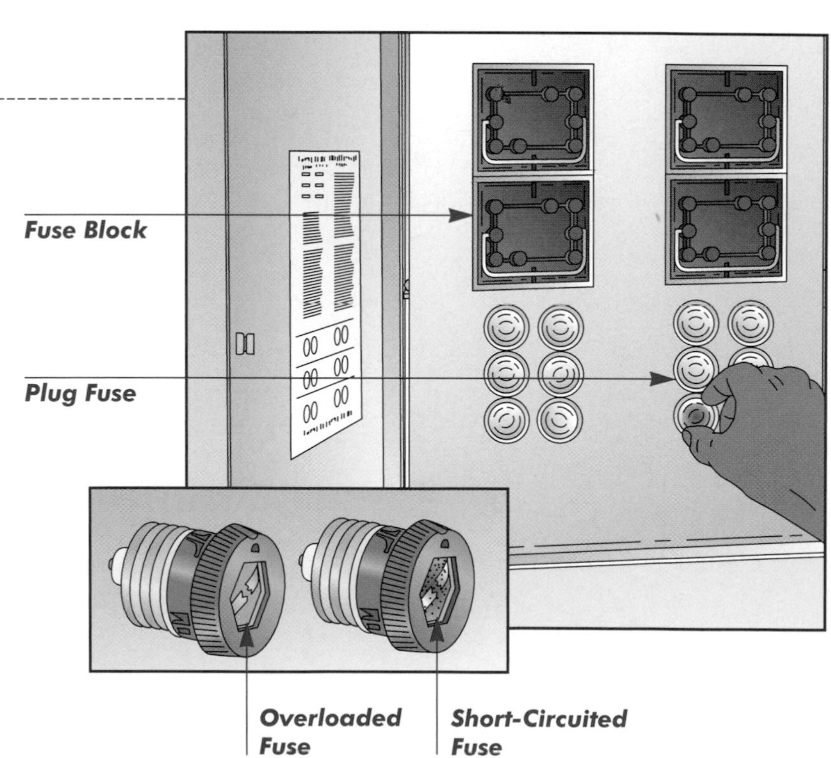

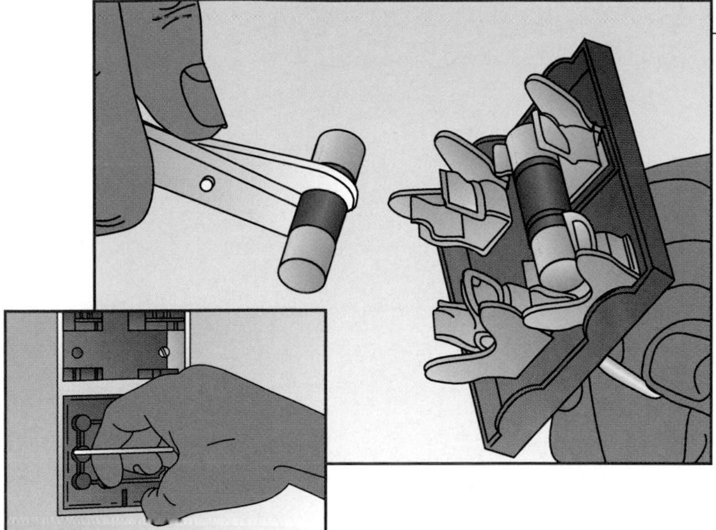

## Removing cartridge fuses

A 240-volt circuit may be controlled by cartridge fuses housed either in fuse blocks or in a separate panel. Before removing a fuse block or cartridge fuse, turn off the appliance it serves.

- To remove a cartridge fuse housed in a fuse block, pull the fuse block out of the service panel by the handle *(inset).*
- Remove the fuse from the spring clips with a fuse puller *(left).* Don't touch metal ends—they may have overheated.
- For cartridges in a separate panel, simply remove them with the fuse puller.

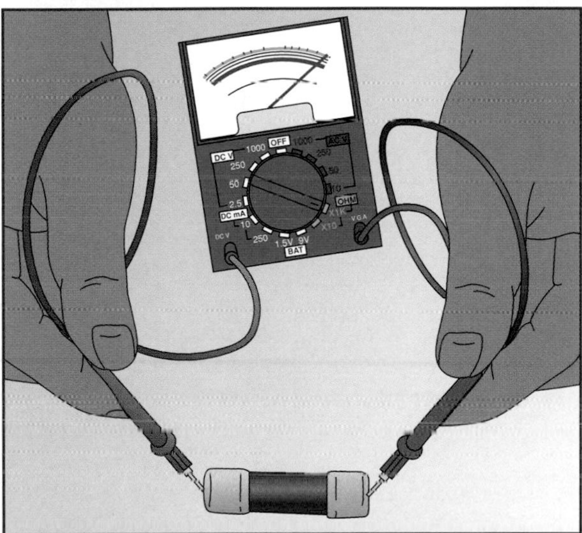

## Testing a cartridge fuse

- Set a multitester *(page 20)* to RX1, and touch the probes to the fuse's metal ends *(left).* A reading of 0 ohms indicates that the fuse is good; an infinity reading indicates that the fuse is blown.
- If the fuse is good, reinstall it in the service panel, inserting it into the spring clips with the fuse puller.
- If the fuse is blown, replace it with a new fuse of identical design and amperage.

## FUSE TYPES AND CAPACITIES

Replace blown fuses with fuses of the same type and amperage—never higher. Common types are shown at right. Standard plug fuses come in 15-, 20-, and 30-amp versions. The metal strip inside a time-delay fuse withstands the momentary power surge created when an appliance motor starts but blows if there is a sustained overload or a short circuit. Type-S fuses fit an adapter, which guards against installation of a higher-amperage fuse. Ferruled-cartridge fuses, rated up to 60 amps, are used to protect circuits for large appliances. Knife-blade cartridge fuses—rated more than 60 amps—are used to protect the house electrical system.

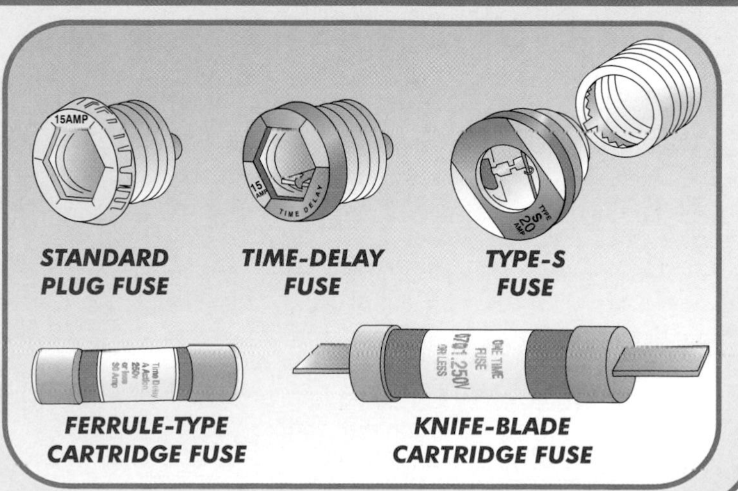

# Mapping Your Home's Electrical System

## Tracing Each Circuit

Having a map of the circuits in your house can speed electrical repairs.

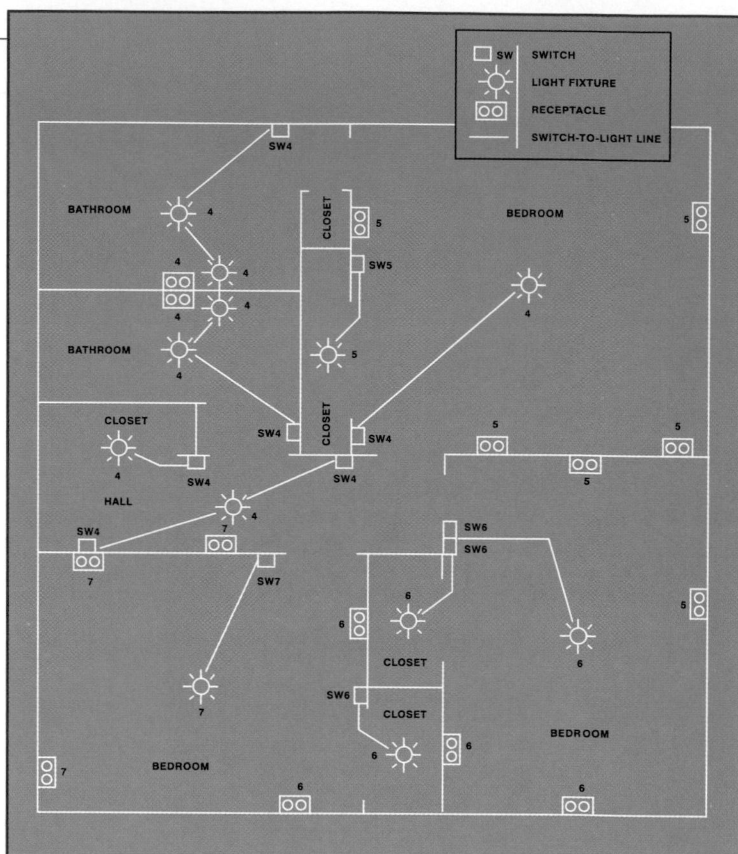


- Sketch a plan of each floor of the house; then mark the location of all outlets, switches, and light fixtures, using standard architectural symbols *(right).*
- Turn off all switches and appliances.
- At the service panel, stick a numbered label next to each circuit breaker or fuse.
- Turn off power to circuit No. 1 by shutting off the breaker or removing the fuse *(page 10).* Go through the house and find the switches, outlets, and fixtures that are not getting power. (Plug in a lamp to test the receptacles.) On the plan, write the circuit number beside each one.
- Repeat the process for each circuit breaker or fuse in the service panel.

## SAFE RESPONSES TO ELECTRICAL EMERGENCIES

A person in contact with live current may appear stuck to the source. Don't touch the individual. Instead, shut off power to the house *(pages 10–11),* or unplug the appliance or lamp. If you can't cut power immediately, use a broomstick or other wooden object to knock the person free *(near right).*

If a lamp or appliance sparks, shocks you, feels hot, or is smoking or aflame, pull its plug. Use a thick, dry towel or heavy-duty work glove to protect your hand *(far right, top).*

Should a switch, outlet, or fixture snap, crackle, spark, smoke, or catch fire, immediately kill power to the entire house.

Never touch a burning or sparking switch to turn it off. Stand away from it and flip off the toggle with a wooden spoon *(far right, bottom).*

# Calculating the Load on a Circuit

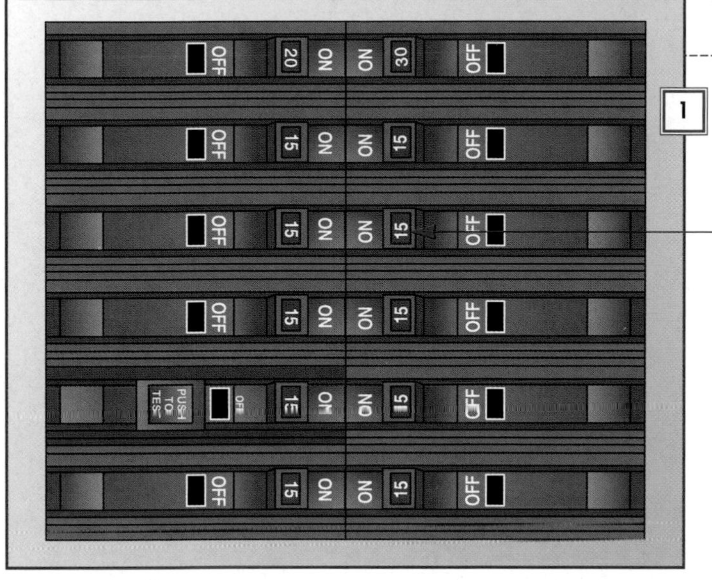


## 1. Determining circuit capacity

If breakers trip or fuses fail persistently—or if you plan to add another load to a circuit—compare the total load on the circuit with its designed capacity, expressed in amperes or amps. To do so:

- Find the amperage rating on the breaker or fuse that controls the circuit in question. On circuit breakers, the amperage rating is embossed on the tip of the toggle switch *(left)*. Plug fuses are labeled and color-coded according to amperage rating *(page 11)*.

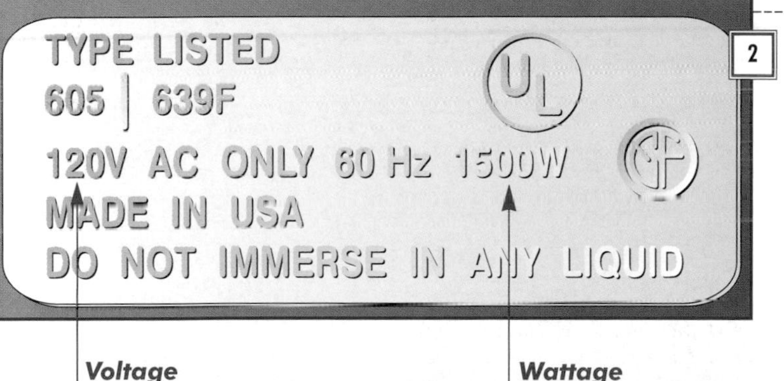


## 2. Calculating the load

- Look for an Underwriters Laboratory label on each appliance *(left)*. The label specifies how much voltage the appliance requires and the power it consumes in watts.
- Add the wattages of all the appliances and light bulbs on the circuit.
- Use the wattage table here to compare the total load to the amperage of the breaker or fuse. For example, a 15-amp circuit breaker or fuse on a 120-volt circuit has a capacity of 1,800 watts.
- If the total load exceeds the capacity, disconnect some loads from the circuit, or consult an electrician.

**How Many Watts Can a Circuit Handle?**

| BREAKER/FUSE RATING | TYPE OF CIRCUIT | |
|---|---|---|
| | 120-VOLT | 240-VOLT |
| 15 AMPS | 1,800 WATTS | 3,600 WATTS |
| 20 AMPS | 2,400 WATTS | 4,800 WATTS |
| 30 AMPS | 3,600 WATTS | 7,200 WATTS |

# FIX IT: Lamp Repairs

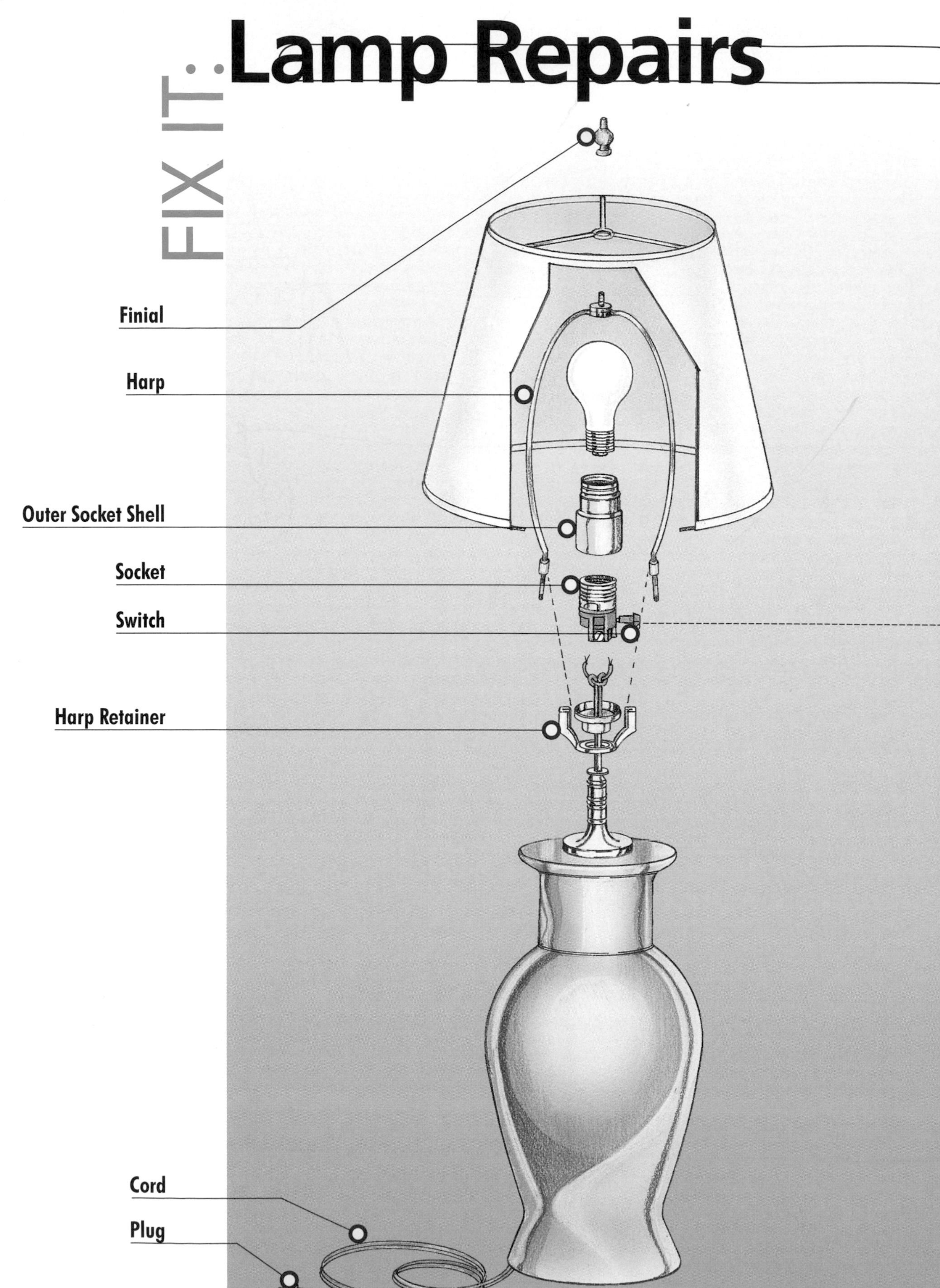

# Chapter 2

## Contents



## How They Work

**In its major components, the table lamp shown at left is virtually identical to most other such lamps, regardless of size, shape, or type of bulb.**

**Every lamp has a cord that runs from a plug at one end, through the lamp, to a socket at the other. In this lamp, the socket accepts an ordinary incandescent light bulb; other sockets accept fluorescent tubes or halogen bulbs.**

**A switch in the socket of this lamp turns it on and off; but switches can also be built into the base of a lamp, into the cord, or in many desk lamps, into the lamp shade. In the typical table lamp shown here, the shade is secured by a finial to a harp, which curves around the bulb.**

# Troubleshooting

| Problem | Solution |
| --- | --- |
| • Bulb flickers or won't light | Make sure lamp is plugged in • Tighten or replace bulb • Test switch and socket **19** • Test plug and cord **30** • Rewire the lamp **30** • |
| • Lamp blows fuse or trips breaker | Reset the breaker or replace the fuse **10** • Make sure circuit isn't overloaded **13** • Test for shorts in the lamp cord **30** • Test plug and socket **19** • Rewire the lamp **30** • |
| • Bulb burns out too quickly | Install long-life bulb or bulb extender • |
| • Shock when changing bulb | Replace socket **21** • |
| • Shock when plugging or unplugging lamp | Test plug and cord **30** • Replace plug **26–29** • Rewire the lamp **30–35** • |
| • Halogen lamp does not light | Replace bulb **35** • Inspect wire at plug and lamp base • Test socket and fuse **35** • |
| • Fluorescent lamp does not light | Replace bulb **36** • Inspect wire at plug and lamp base **36** • Test socket and fuse **36–37** • Replace ballast **37** • |

# Before You Start

## Look for simple problems first, then follow a thorough diagnostic procedure

### LAMP-REPAIR LOGIC

If a lamp doesn't work, first make sure the bulb hasn't burnt out, that the lamp is plugged in, and that the fuse or breaker controlling the circuit hasn't blown or tripped *(page 10)*. If you find nothing amiss, test the socket next, then look at the plug, and test for continuity in the cord. Never try to repair a worn lamp cord; install a new one that has wires of the same diameter—or gauge—as the original. And while you're at it, replace the plug, too.

### Before You Start Tips:

- Lamp cord is made of fine wire strands. Twist them together, when making connections, so there are no stray ends.
- Buy polarized replacement plugs, which have one narrow and one wide prong. Wired correctly, this kind of plug assures that exposed, metal parts of the lamp will be disconnected from the power when the lamp is off.
- Twisting on a wire cap is the best way to join wires. Check the package label to be sure you have the right size for the number and diameter of the wires at the connection.

- Screwdrivers
- Multitester
- Wire stripper
- Utility knife
- Paring knife
- Long-nose pliers
- Multipurpose electrical tool

- Lamp cord
- Plugs
- Wire caps
- Electrical tape

Always unplug a lamp before working on it.

# Testing and Replacing a Socket

## 1. Cleaning and Adjusting the Contact Tab

To make a good electrical connection, the socket's contact tab must be clean, and it must fit tightly against the base of the bulb.

• With a flat-tip screwdriver, scratch any dirt off the surface of the tab. If it is corroded or broken, remove the socket and replace it with a new one as shown on these pages.

• Check that the tab angles slightly upward from the base of the socket. If not, pry it gently upward *(right).*

• If the contact tab snaps off as you pry it—or if the bulb still doesn't light afterward—replace the socket.

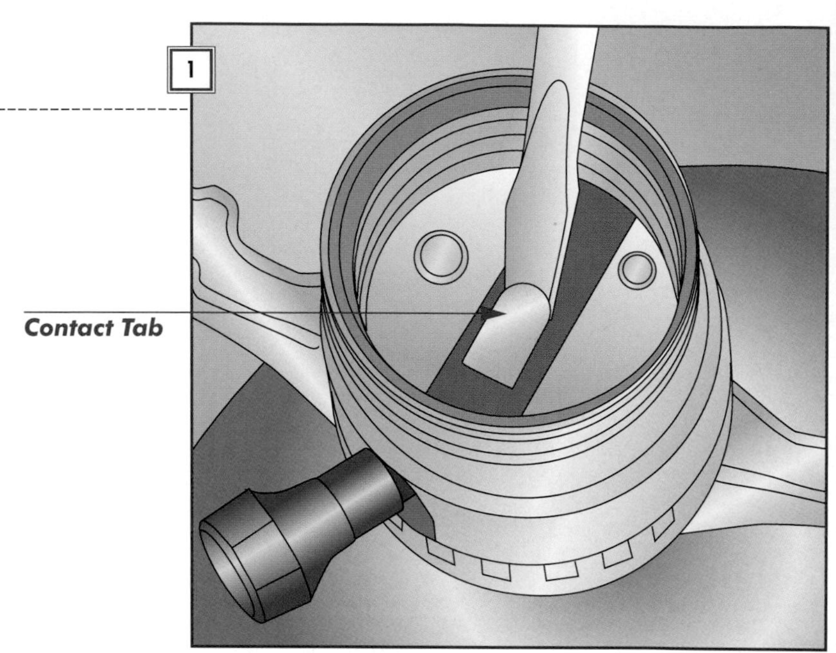


## 2. Removing the Outer Shell

• On the outer shell, look for the word PRESS *(not shown).* Push hard there with your thumb *(right).* Squeeze and lift the shell from the socket cap. You may have to wiggle the shell slightly, but don't twist it.

• Slip off the cardboard insulating sleeve if it doesn't come off with the shell. Examine the sleeve. If it is damaged, replace the socket.

• Disconnect the socket by loosening the two terminal screws and removing the wires *(inset).*

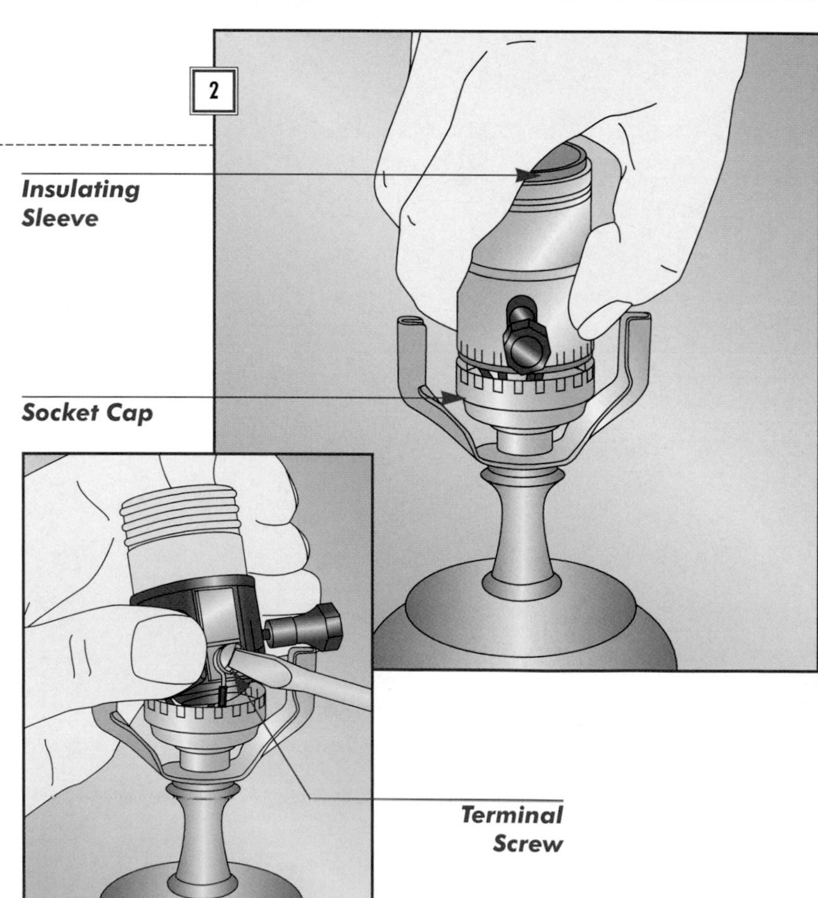

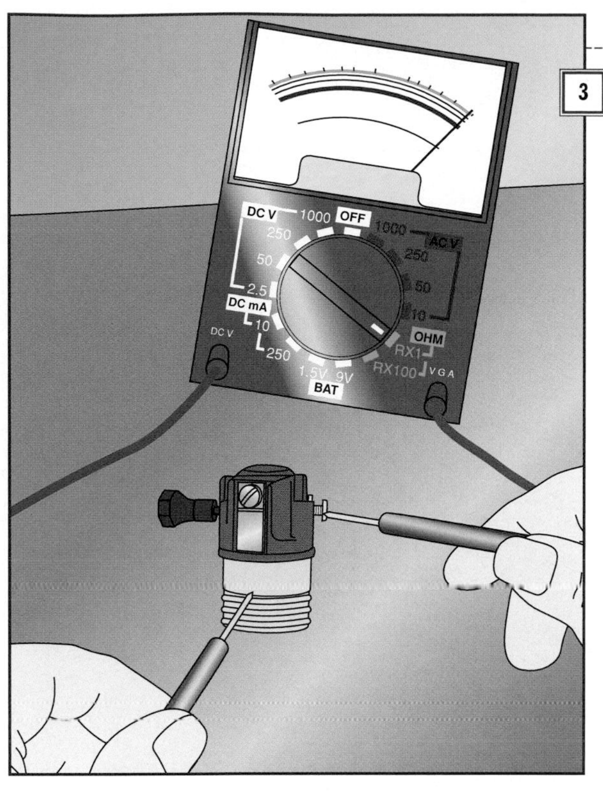


## 3. Testing the Socket

• Set a multitester to RX1 for a continuity test *(page 20)*.

• Touch one probe to the silver (neutral) terminal screw and the other to the socket's threaded metal base *(left)*. The meter should read 0 ohms, indicating full continuity.

• If you don't get a 0 reading, the socket is faulty; replace it. Otherwise, continue by testing the switch *(Step 4; also pages 23 and 25 for other types of switches)*.

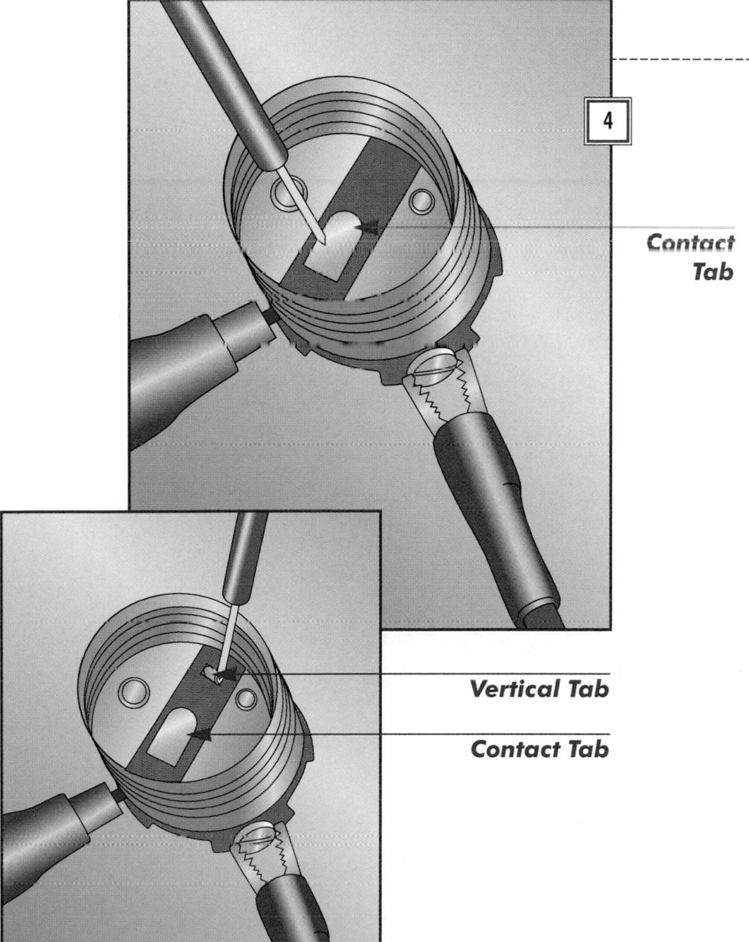


## 4. Testing the Switch

A properly working switch will show continuity only when it's in the ON position.

• To test a two-position ON/OFF switch *(left)*, place the the leads of a multitester on the brass terminal and the socket contact tab. The meter should read 0 ohms with the switch in ON position, infinity in OFF.

• To test a three-way switch *(inset),* place the probes against the brass screw terminal and either the small vertical tab in the socket base or the contact tab. Where you detect continuity depends on the switch position. In one of its ON positions, a properly working switch shows continuity with the probe touching the vertical tab; in another ON position, when it touches the contact tab; in the third ON position, when it touches either tab. When turned off, the switch shows no continuity.

• If the switch is faulty, replace the socket and socket cap *(page 21)*. If tests indicate that the switch and socket are good, test the cord and plug *(page 30, Step 1)*.

## USING A MULTITESTER

A multitester is an inexpensive but invaluable tool for diagnosing electrical problems. Multitesters come in two styles. The analog type, shown here, has a needle indicator and printed scales that correspond to the settings of a selector dial used to choose the test mode. Digital multitesters give numerical readouts according to the test mode selected. All multitesters have probes connected to positive and negative leads. Alligator clips can be slipped over the probes and clipped to terminals when convenient.

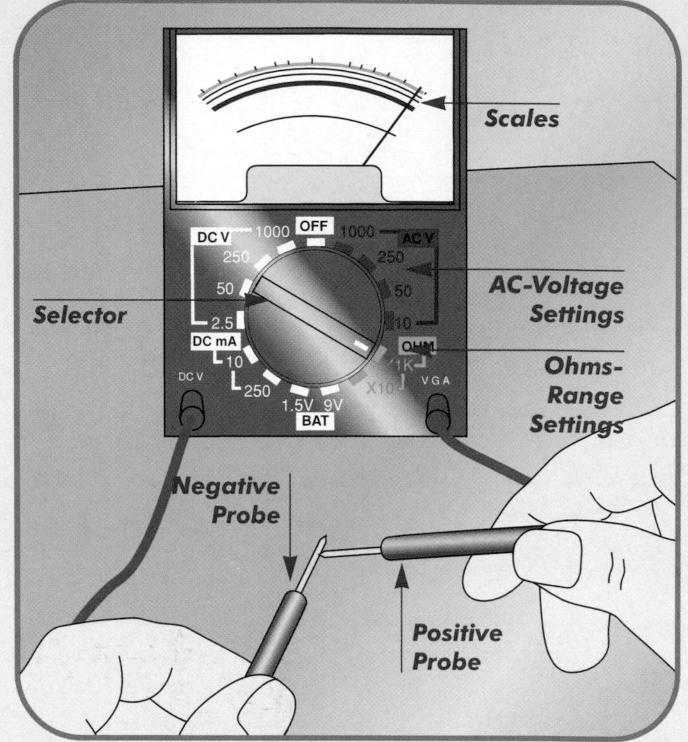


**ZEROING THE NEEDLE.** Set the selector dial to RX1 or any other ohms-range setting. Touch the probes together *(right).* On analog multitesters, the needle should sweep from left to right. Turn the ZERO OHMS adjustment dial until the needle rests directly over 0. Digital meters do not require zeroing.

**CHECKING CONTINUITY.** A continuity test determines whether a circuit is closed complete or open. Turn off power to the circuit and disconnect the wires or component to be tested. Set the selector to RX1 or any ohms setting, and touch the probes to two different terminals or to the ends of a wire.

The illustration at center right shows a continuity test on a switch. With the switch ON and probes touching the two terminals, the needle should indicate 0 resistance—a closed circuit. An infinity reading indicates an open one.

**TESTING FOR VOLTAGE.** Voltage checks reveal whether a circuit is energized and how much voltage is present. In the illustration at bottom right, the probes are inserted in the slots of an outlet. A reading other than 0 indicates that voltage is present and that it would be unsafe to touch or disconnect the component without cutting power. When testing whether voltage flows within the correct range, for a 120-volt circuit, look for a voltage reading between 108 volts and 132 volts; for a 240-volt circuit, a reading between 216 volts and 264 volts.

**CAUTION:** *When testing for voltage, hold multitester probes by their plastic grips to avoid possible electrocution.*

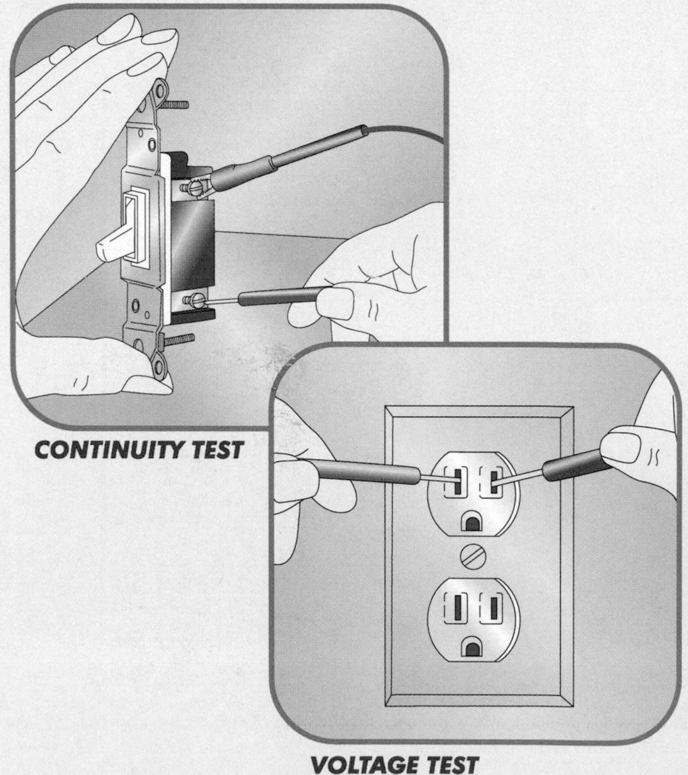

**CONTINUITY TEST**

**VOLTAGE TEST**

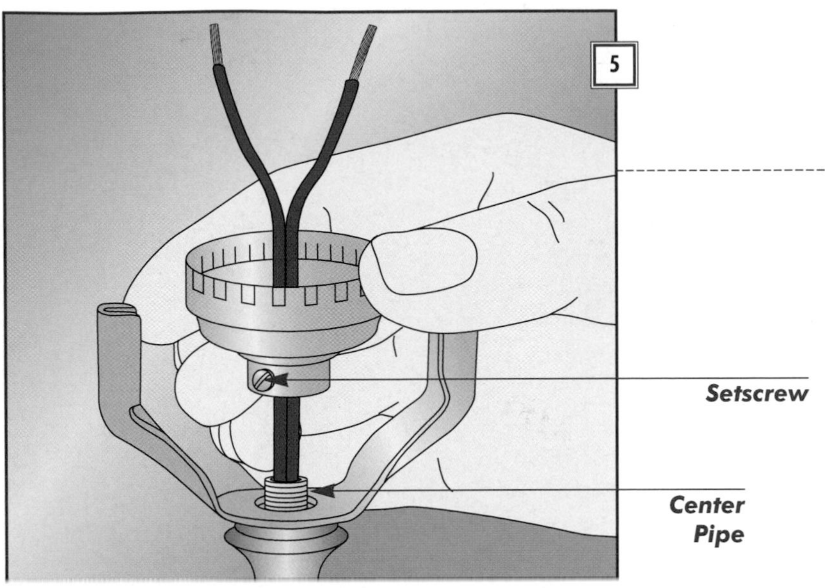


## 5. Removing the socket cap

• If the lamp cord is knotted inside the socket cap, untie it.

• Loosen the setscrew, if any, at the base of the cap, and unscrew the socket cap *(left).* If the socket cap holds the lamp body to the center pipe, support the lamp so that it doesn't come apart.

• Inspect the cord. If it's in poor condition, rewire the lamp *(pages 30–35).*

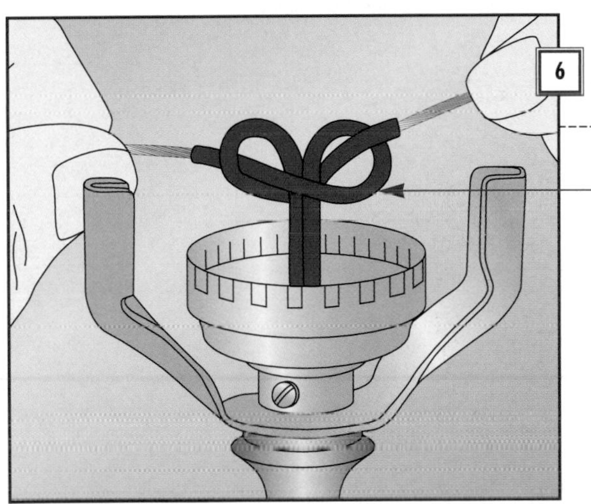


## 6. Installing a new socket cap

• Thread the lamp cord through the cap, then screw the cap onto the center pipe. Tighten the setscrew, if present.

• If the cord is new, part the wires and strip the ends *(page 29).* Tie the ends of the parted cord in an Underwriters knot *(left).*

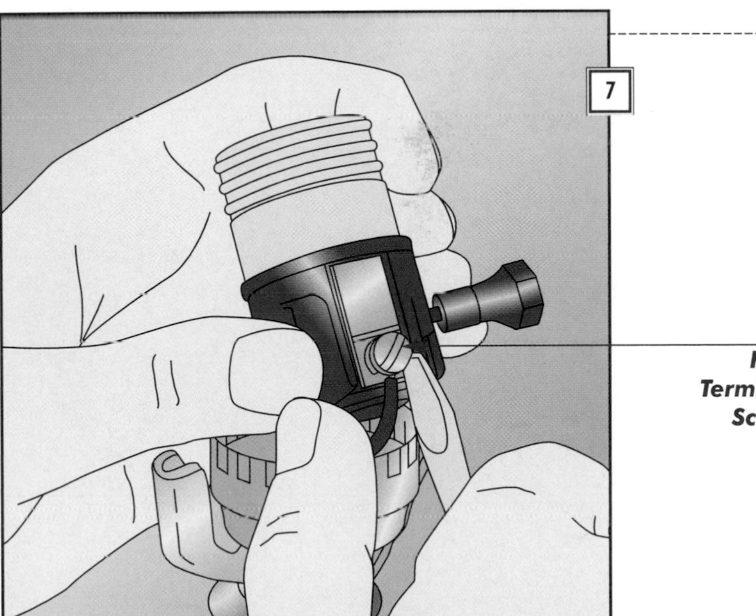


## 7. Wiring the new socket

One of the wires of a lamp cord has a ridge molded into the insulation to mark it as the neutral conductor; the hot-conductor insulation is smooth. These features help to assure correct connections to sockets, switches, and plugs.

• Connect the ridged wire to the silver socket terminal and the smooth one to the brass terminal. To do so, twist the wire strands clockwise so there are no frayed ends, and form a hook at the end of each wire. Use a screwdriver to help wrap each hook around its terminal screw *(left).*

• Tighten the screws, making sure there are no stray wire strands.

## 8. Completing the installation

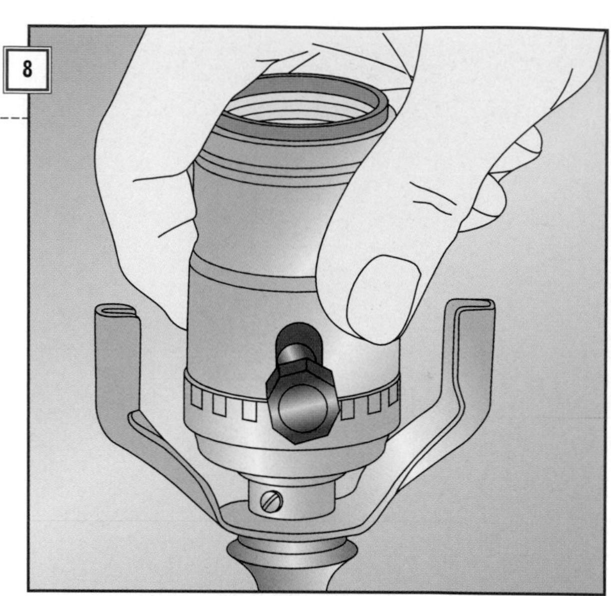


• Slip the insulating sleeve and outer shell onto the socket. Fit the notched opening over the switch, then push the shell into the rim of the cap, snapping it into place *(right).*

• Screw in the bulb, then plug in the lamp and turn it on.

• If the bulb lights, reassemble the lamp as needed. If the lamp doesn't light, test the plug and cord *(page 30).*

## INCANDESCENT LAMP SOCKETS

Although all sockets in household lamps and light fixtures are similar, there are many designs for different purposes. When replacing a socket, take the old one with you to the store to find an exact match.

1. Medium-base lamp socket: can be made of metal or plastic; it may have a simple ON/OFF switch or a three-way switch.

2. Porcelain socket: usually attaches to a threaded tube.

3. Plastic-fixture socket: may have preattached leads and usually attaches to a fixture with an external mounting.

4. Two-part socket: separates into two components that are fitted into a socket hole from each side and screwed together.

5. Outdoor socket: designed to resist corrosion; may be made of porcelain, plastic,or rubber.

6. Low-voltage socket: designed for low-voltage bulbs; may have a threaded or bayonet base.

# Dealing with a Switch in a Lamp Base

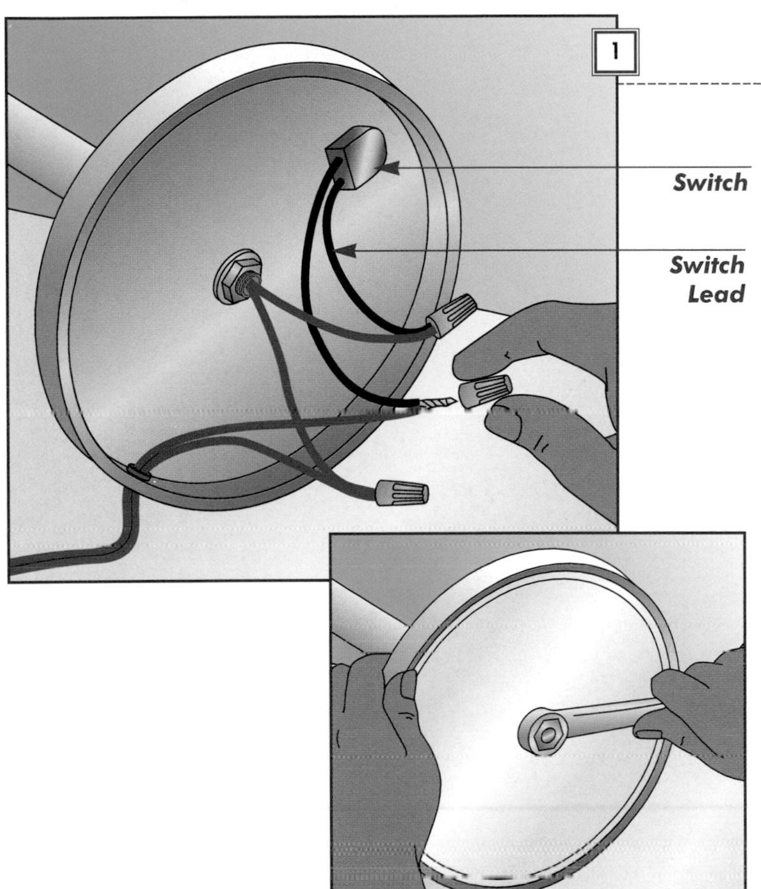


## 1. Uncovering the Switch Leads

Some lamps have a switch mounted in the base instead of in the socket or cord. If the bulb flickers or if the switch feels loose on such a lamp, take out the switch and test it.

- Unplug the lamp, set it on its side, and peel off the protective cover from the base. If the lamp has a bottom plate, remove the locknut holding it in place *(inset),* then pull off the plate.

- Disconnect the switch leads *(left).*

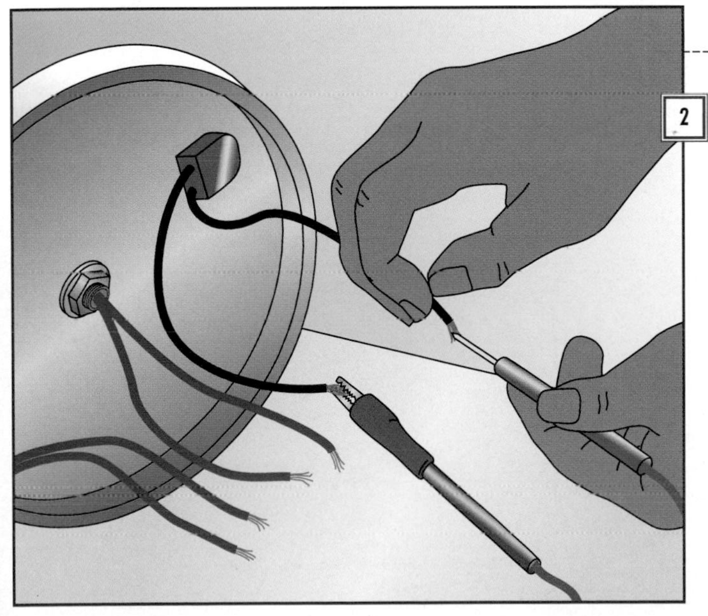


## 2. Testing the Switch

- Set a multitester to RX1 *(page 20),* and touch the multitester probes to the ends of the switch leads *(left).* Note the reading on the multitester scale—either 0, which indicates continuity,or infinity, which indicates an open circuit.

- Push or turn the switch once and note the multitester reading again. It should be the opposite of the result from the first test. If not, replace the switch *(Steps 3–4).*

## 3. Removing the switch

- Unscrew the switch-retaining ring, then pull the switch out of the base *(right).*
- Take the switch to a hardware store or electrical supplier, and get a compatible replacement.

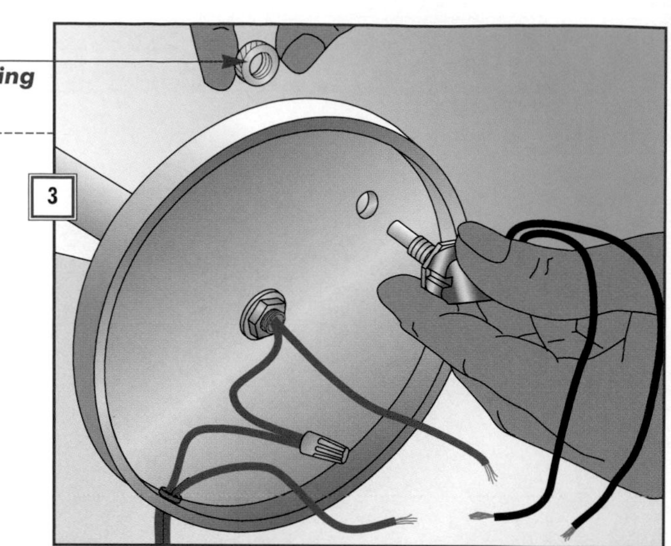


## 4. Installing the new switch

- Place the new switch in the lamp base, and tighten the retaining ring.
- Twist each switch lead together with one of the wires to which the old switch was connected *(right),* and secure each connection with a wire cap.
- If you removed a bottom plate, reattach it with the locknut.

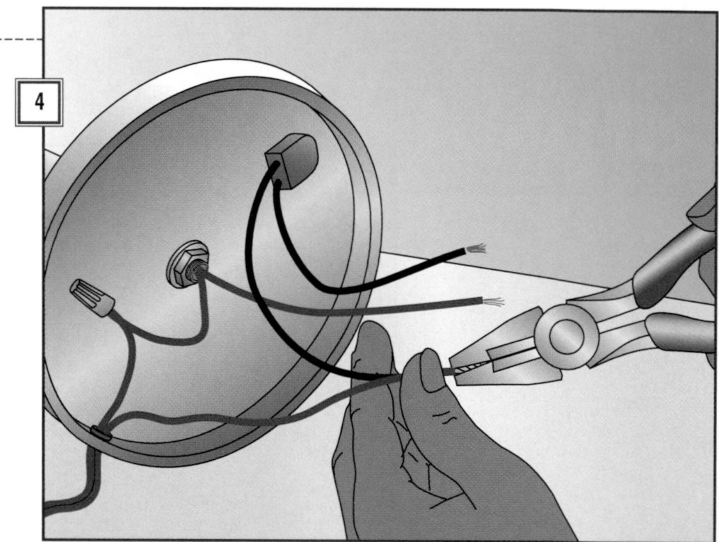

## Ron's TRADE SECRETS

### Dealing with a broken bulb

I've got a couple of favorite tricks for unscrewing a broken light bulb: First, I unplug the lamp. Next, I hold a pair of long-nose pliers so that I can spread the jaws with my fingers and thumb to press against the inner walls of the bulb's threaded base. Then I turn the pliers to unscrew the base from the socket. If you have difficulty spreading the pliers with one hand, use two—and get a helper to hold the socket.

If this method fails, I work a small screwdriver between the socket wall and bulb base to bend it slightly inward. That gives me something to grab and turn with the tips of the pliers.

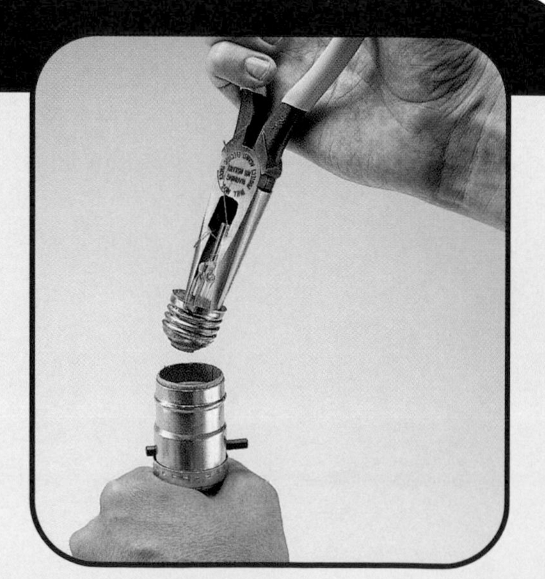

# How to Test and Replace Cord Switches

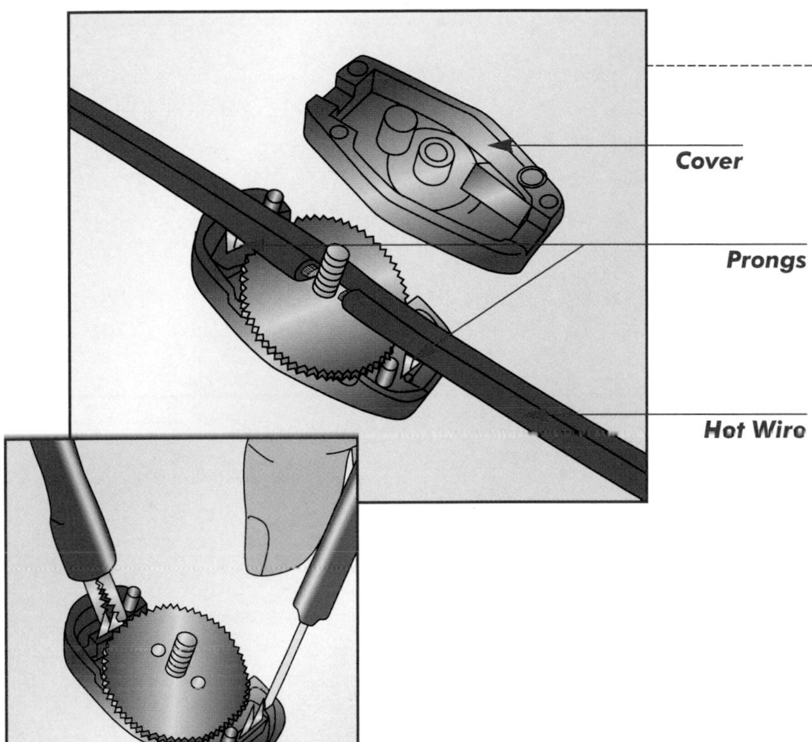


### Lamp-cord switches

• Unplug the lamp, remove the switch cover *(left),* and lift the cord out of the switch.

• Set a multitester to RX1, and touch the probes to the switch terminals to test for continuity *(inset).* The meter should indicate continuity with the switch in the ON position, and no continuity after being clicked OFF. If your results are different, replace the switch.

• Check the lamp cord. If it is sound, install the new switch in the existing cord. If not, replace the cord *(pages 30–35).*

• Notch the hot wire (smooth insulation) of a new cord to fit around the switch's central screw. Make sure the switch's prongs pierce the hot-wire insulation.

• Attach the switch cover.

### Round-cord switches

• Unplug the lamp, remove the screws that hold the switch cover in place, and put it aside *(inset).*

• Set a multitester to RX1, and touch the probes to the switch terminals to test for continuity *(left).* The meter should indicate continuity with the switch in the ON position, and no continuity after being clicked OFF. If your results are different, replace the switch.

• When installing a switch on a new cord, carefully cut the outer insulation to expose the conductors inside without damaging their insulation.

• Cut the black wire, strip the insulation off the ends *(page 29),* and connect them to the terminal screws. Trim excess wire, then attach the switch cover.

# Inspecting a Plug

### EXAMINING THE DESIGN

A glance tells whether a plug's casing is cracked or if the prongs are loose, bent, or corroded. If you find any of these defects, replace the plug. Usually it's best to replace a plug with a similar one. Four common types are shown at right. One prong of a polarized plug is wider than the other so that the neutral wire in a lamp, for example, can't accidentally be connected to the hot side of an outlet.

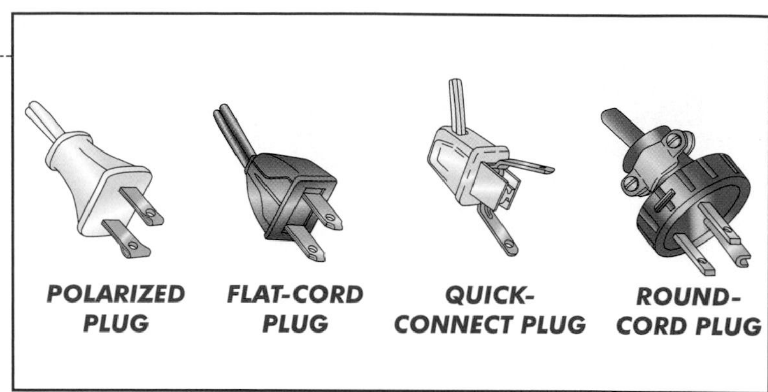


# Replacing a Flat-Cord Plug

### 1. THREADING THE SHELL

- Unplug the lamp and cut off the old plug with wire cutters.
- Separate the core of the new plug from the shell, and slip the cord through the shell *(right).*
- Part the wire and strip about 1/2 inch of insulation from the ends *(page 29).*
- To protect the connections from strain, tie an Underwriters knot with the two wire ends *(page 21).*

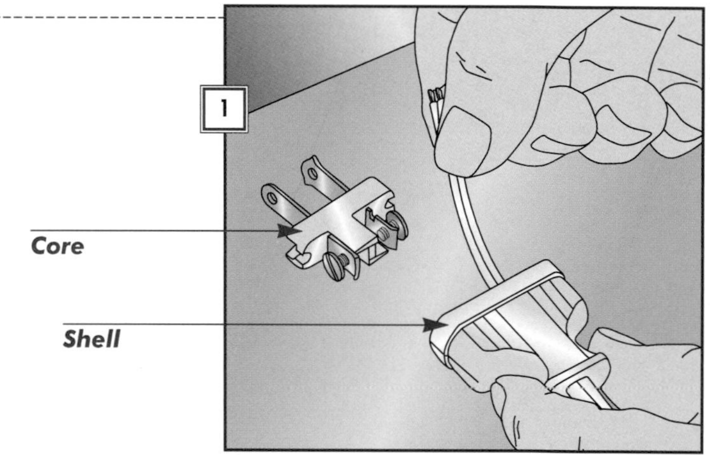


### 2. CONNECTING THE TERMINALS

- Twist the wire strands together and use a screwdriver to hook the end of each wire clockwise around a terminal screw *(right).* The hot wire (smooth insulation) should go to the plug's narrow prong; the neutral wire should go to the wide prong.
- Tighten the terminal screws, making sure there are no stray wire ends.
- Snap the core into the shell and tighten any retaining screws.

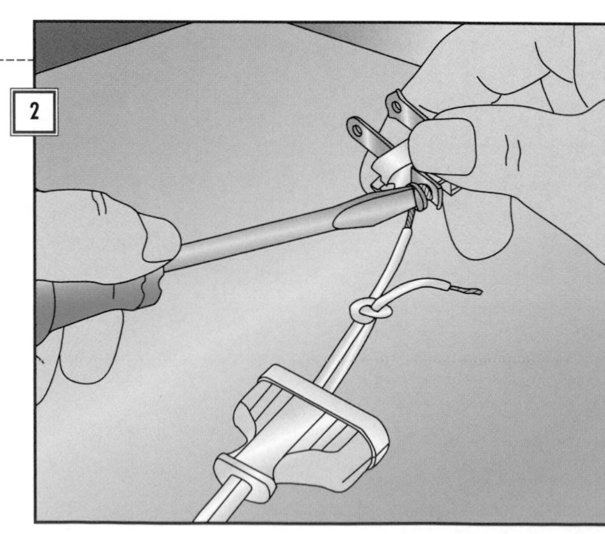

# Quick-Connect Plugs

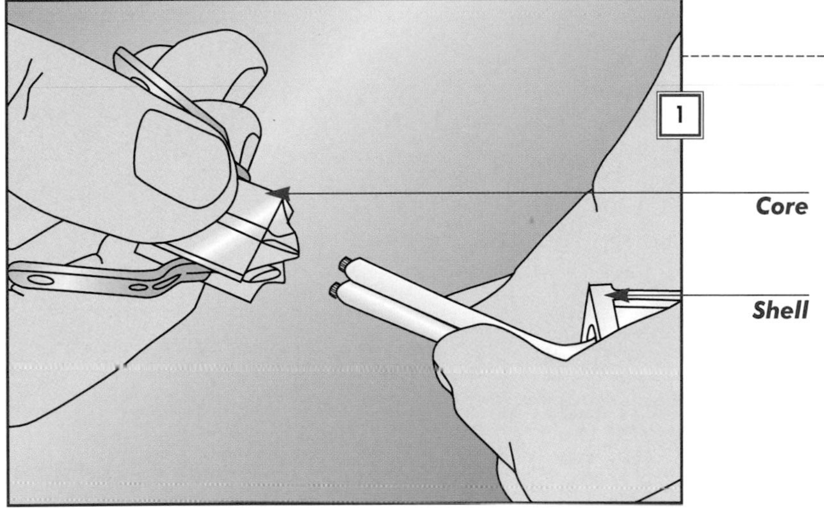


### 1. Inserting the wire into the core

Some quick-connect plugs have a core-and-shell configuration, as shown here. Another common type has no shell; the plug clamps onto the cord and releases with a lever.

- Unplug the lamp, and cut the old plug from the cord.
- If the plug is the core-and-shell type, pull these two pieces apart and thread the cord through the shell of a new plug. Spread the prongs by hand *(left)* and insert the lamp cord, aligning the neutral wire with the wide prong. If the plug is the lever type, lift the lever and insert the cord.

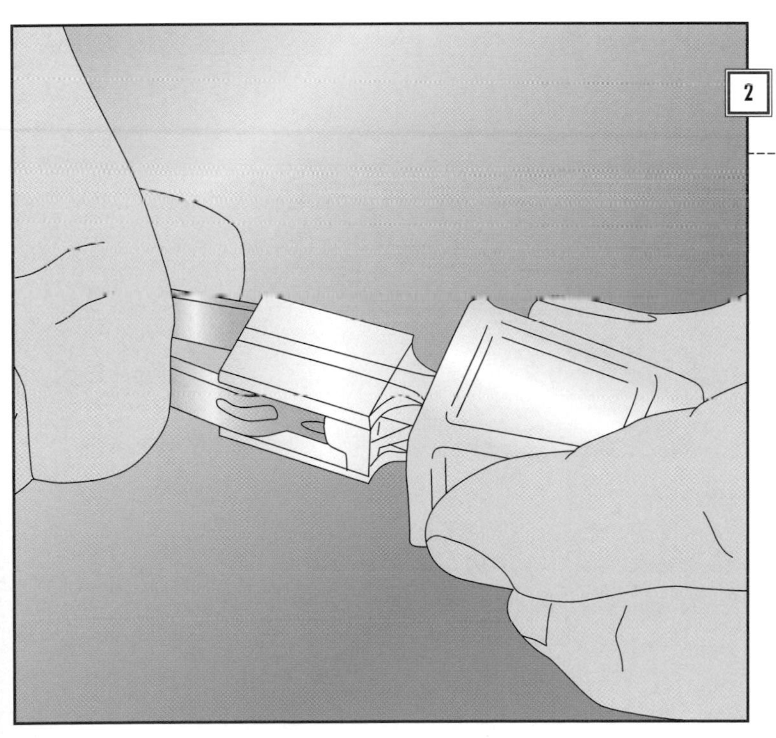


### 2. Making the connection

- With a two-part plug, squeeze the prongs together by hand *(left)*. With a lever-type plug, depress the lever, causing sharp points on the prongs to penetrate the cord insulation and contact the wires inside.
- Slide the shell, if any, over the plug.

# Replacing the Plug on a Round Cord

## 1. Removing the old plug

Round lamp cord is always connected to the plug by screw terminals. If the insulating disk is missing, or if the prongs are bent or broken, replace the plug.

- Unplug the lamp.

- Cut the old plug from the cord with wire cutters *(right)*, and replace it with a new one.

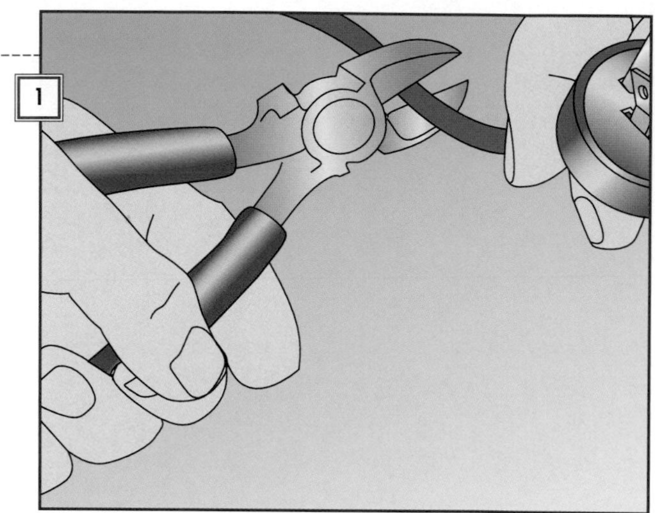


## 2. Threading the shell

- Pry the shell off the new plug's core with a flat-tipped screwdriver.

- Slide the shell onto the cord *(right)*, loosening the plug clamp as necessary.

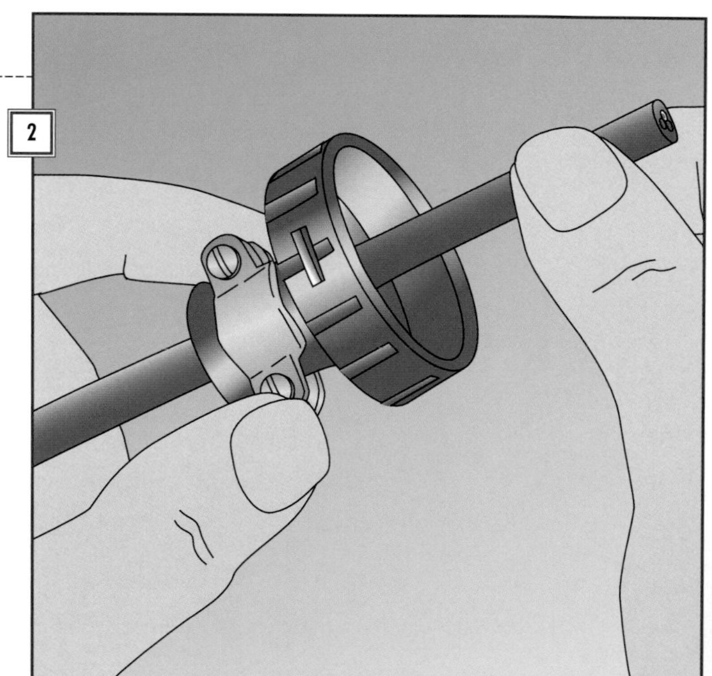

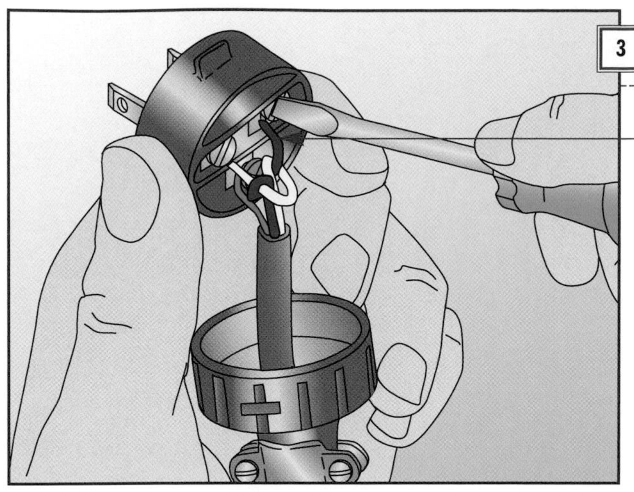


Ground Terminal

### 3. Connecting the wires

• With a utility knife, carefully cut off approximately 1-1/2 inches of the outer insulation to expose the conductors inside without damaging their insulation.

• With a multipurpose tool *(page 110),* strip approximately 1/2 inch of insulation from the white, black, and green wires.

• Hook the white wire around the silver terminal screw and the black wire around the brass screw *(left).* Connect the green wire to the ground terminal.

### 4. Tightening the clamp

• Snap the shell into place over the core.

• Tighten the screws of the plug clamp to secure the plug to the cord *(left).*

## WORKING WITH LAMP CORD

Before attaching a cord to a lamp, you will need to separate the cord's two conductors and remove some insulation from each one. To split the cord, insert the point of a utility-knife blade between the conductors about 1-1/2 inches from the end of the cord *(top right),* then pull the cord past the blade.

Best for stripping insulation is the multipurpose tool *(page 110),* but a utility knife also works well. Place the wire on a flat surface. Taking care not to sever any of the copper strands, shave about 1 inch of insulation from one side of the wire *(bottom right).* Then peel back the insulation and cut it off.

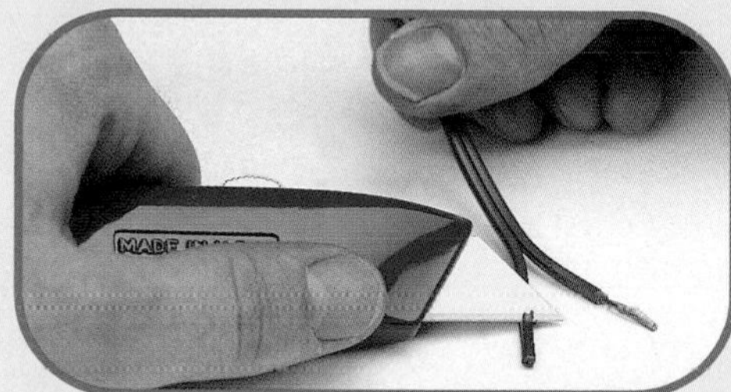

# Rewiring a One-Socket Lamp

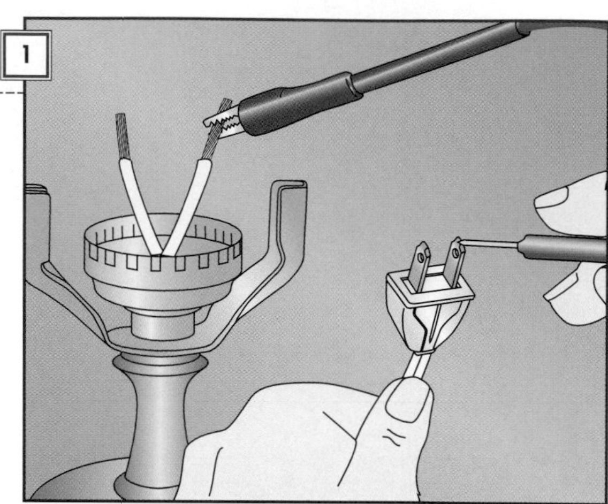


## 1. Testing the Cord and Plug

• Unplug the lamp and remove the socket *(page 18).*

• Set a multitester to RX1, and clip one probe on one wire end *(right).* Place the other probe against one plug prong and then the other. The meter should indicate continuity on one prong only *(page 20).*

• Repeat for the other wire end. If the cord and plug fail either test, rewire the lamp *(Steps 2–3).*

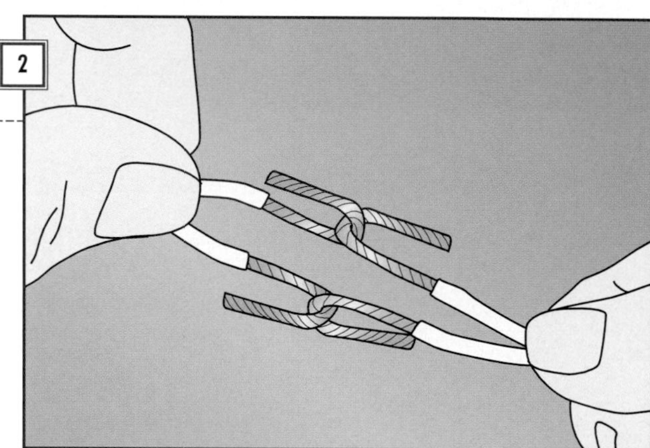


## 2. Tying the Old Cord to the New

• Strip the ends of the new cord *(page 29),* and twist the strands clockwise.

• At the top of the lamp, hook the ends of the old and new cords together *(right).*

• Wrap electrical tape around the connection. If it won't fit through the lamp, unhook one pair of wires and retape them alongside the remaining hooked pair.

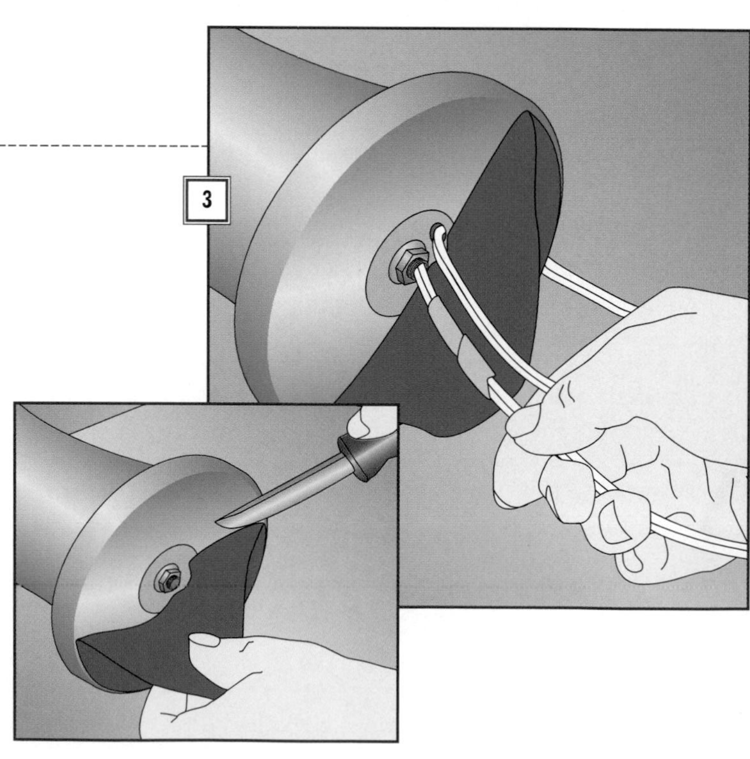


## 3. Fishing the New Cord

• With a paring knife, carefully peel back the base's felt cover far enough to uncover the cord in the lamp base *(inset).*

• Feed the connection into the center pipe at the top of the lamp. From the base, pull on the old cord until the connection appears *(right).* Then pull on the old cord at the edge of the lamp base to bring the new cord through the channel.

• Untape the connection, and discard the old cord.

• Install a new socket *(page 21)* and plug *(pages 26–27).*

# Rewiring a Multi-Socket Lamp

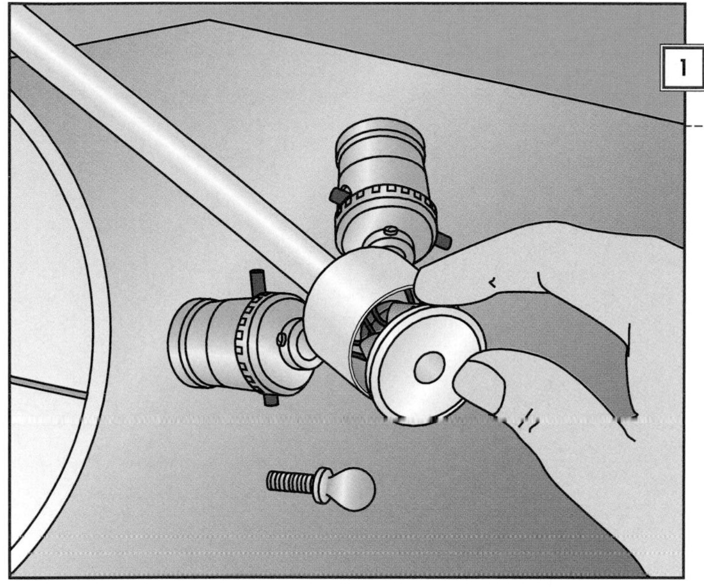


### 1. Exposing the wires

Multiple-socket lamps are represented in the following steps by a lamp having two sockets, but the procedures are the same for lamps with three or even four sockets.

- Unplug the lamp and remove the shade.
- Unscrew the cover at the junction of the sockets *(left)*, then remove the outer shells of the sockets *(page 18)*.
- Disconnect the wires from the socket terminals, and set the sockets aside.

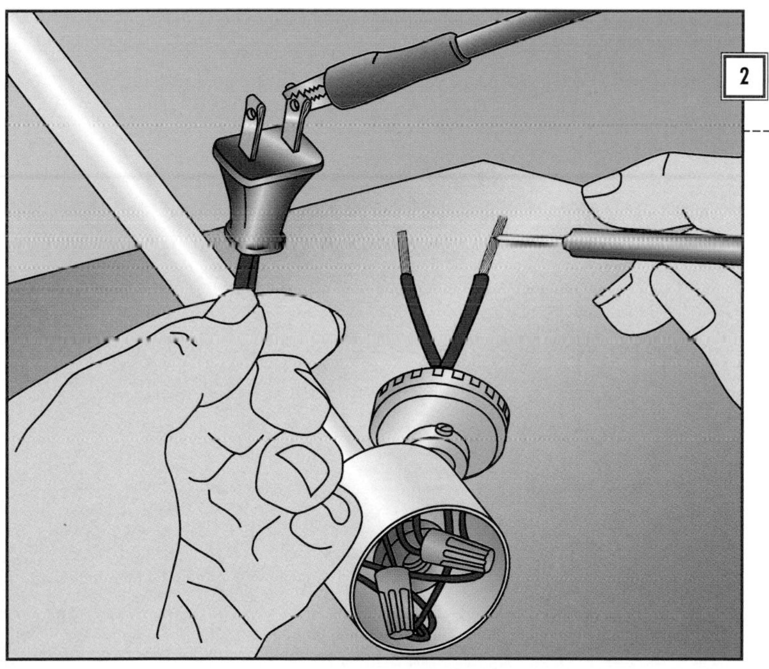


### 2. Testing the wiring

- Set a multitester to RX1, and clip one probe to one plug prong. Touch the other probe first to one wire end, then to the other. The meter should indicate continuity *(page 20)* on one wire only. Repeat the test on the other plug prong and both wires.
- Do the test again with the plug and both wires to the second socket.
- If the cord and plug fail either test, rewire the lamp *(Steps 3–6)*.

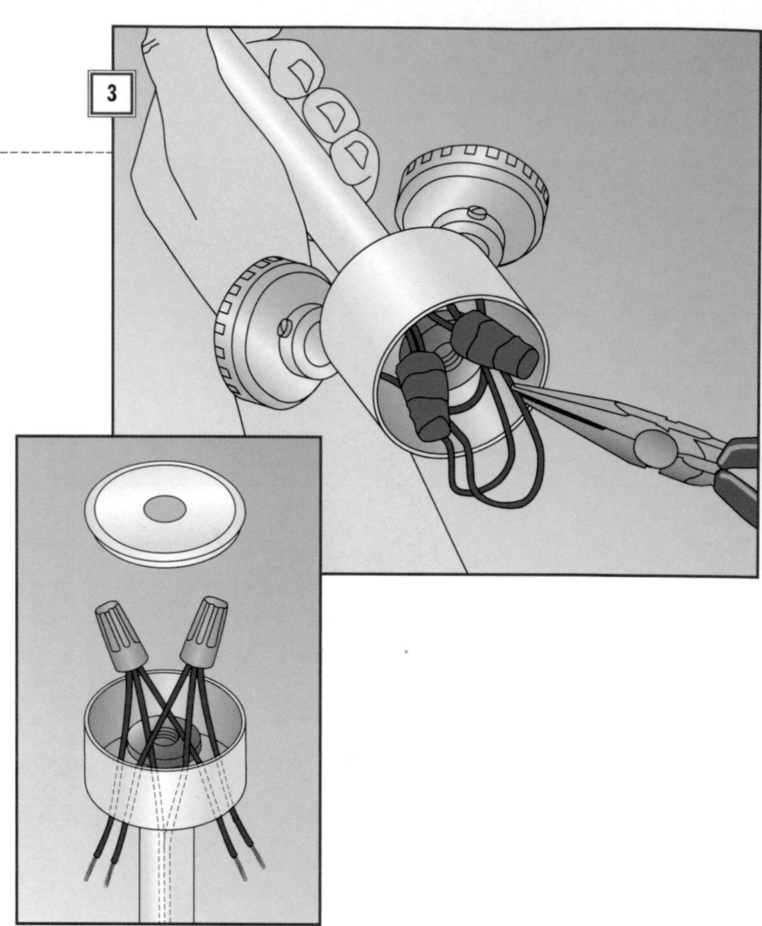


## 3. Disconnecting the wires

• Pull the socket wires out of the socket caps with long-nose pliers *(right).*

• Note that one wire to each socket is connected to each of the lamp cord wires *(inset).* Remove any electrical tape, twist off the wire caps, and disconnect the wires.

• Cut a new lamp cord long enough to extend at least 6 feet from the base of the lamp, and fish it through the lamp's center pipe *(page 30).*

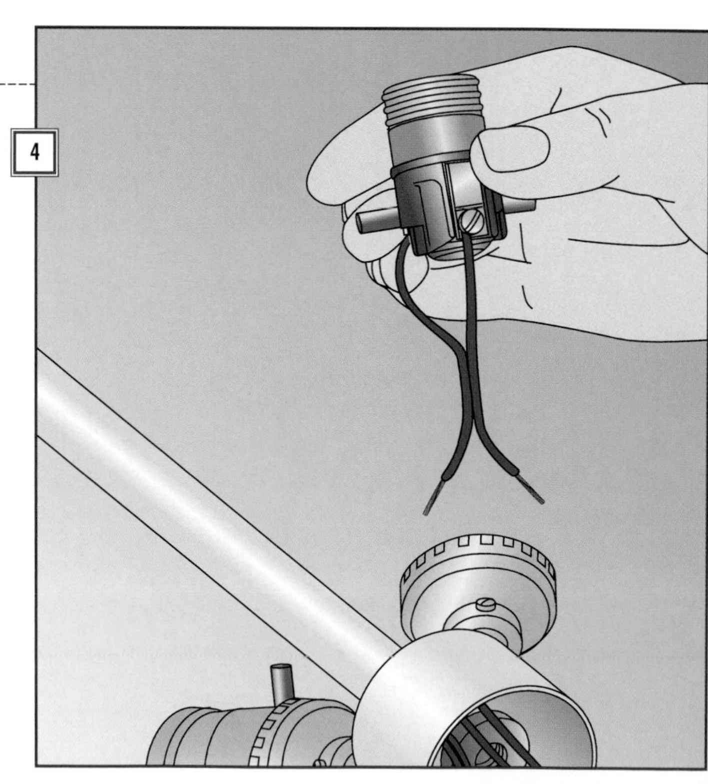


## 4. Rewiring the socket

• For each socket, cut a piece of lamp cord the same length as the old socket wires, then part and strip both ends of both socket wires, as well as the upper end of a new lamp cord *(page 29).*

• Connect the neutral wire—the one with the ridge in its insulation—to the silver socket terminal and the other wire to the brass socket terminal.

• Place the socket in the lamp, feeding the wires through the socket cap *(right).*

• Reinstall the socket's insulating sleeve and shell, pushing on the shell until it snaps into place *(page 22).*

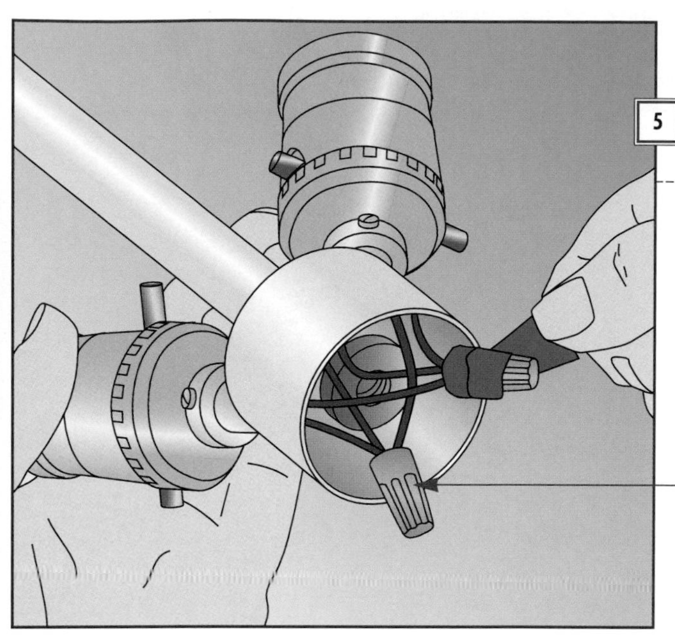


## 5. Connecting the socket wires to the lamp cord

• Twist the cord's neutral wire together with the neutral wires from each socket, and the cord's unridged wire with the unridged wires from each socket.

• Secure each connection with a wire cap. As an extra measure of security, wrap the wire caps with electrical tape *(left).*

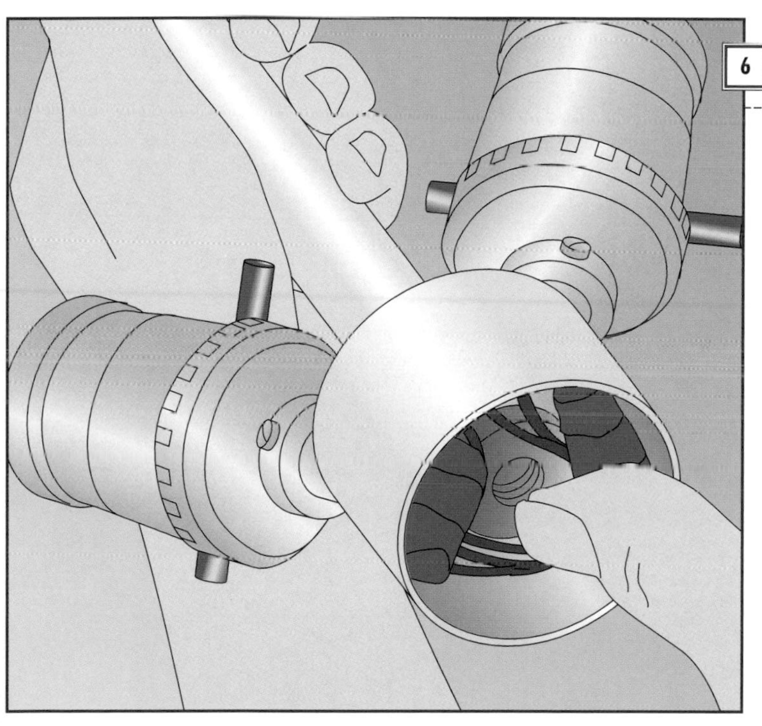


## 6. Reassembling the lamp

• Fold the wires back into the top of the lamp and replace the cover *(left).*

• Connect a new plug to the lamp cord *(pages 26–29),* and reinstall the shade.

# Rewiring an Incandescent Desk Lamp

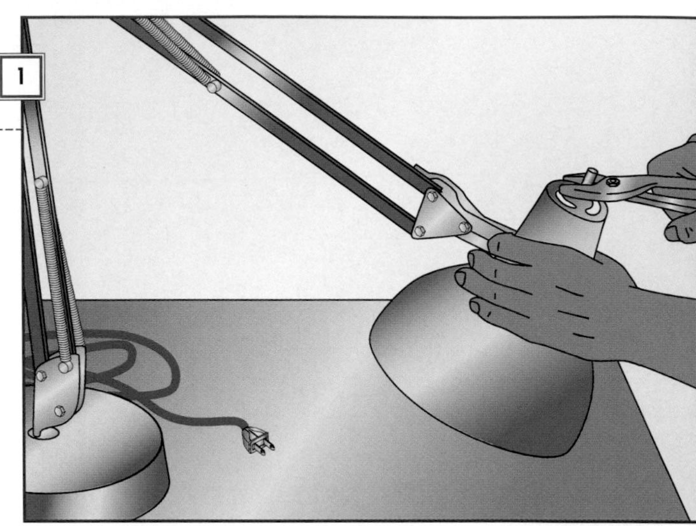


### 1. Disconnecting the socket-retaining ring

• If your lamp has a switch at the top of the shade, use pliers to unscrew the socket-retaining ring *(right).* The method for getting at the socket may vary for lamps with a switch in the base.

• Push the cord into the lamp at the base to gain some slack. Pull the socket out of its ceramic or plastic insulating sleeve.

• Use a screwdriver to loosen the terminals, and disconnect the wires.

• With a multitester, test the cord for continuity *(pages 20 and 30).* If it fails, rewire the lamp *(Steps 2–3).*

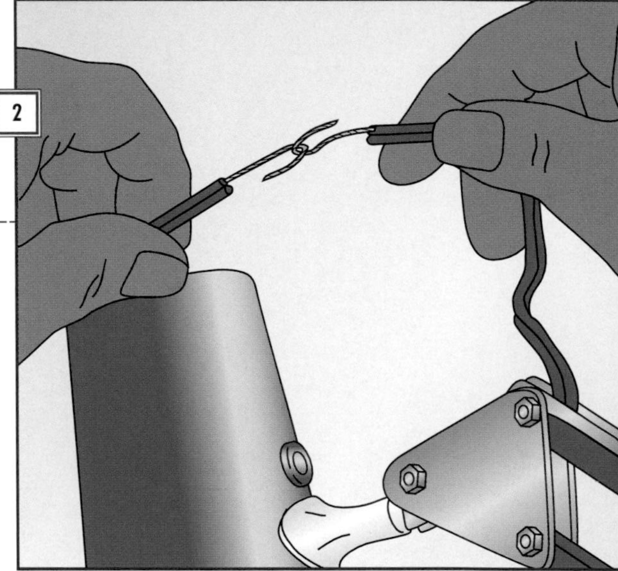


### 2. Tying the old cord to the new

• Pull the cord through the hole in the lamp shade so that its wire ends are exposed at the top of the upper arm.

• To make a joint that's thin enough to fit through the channel of the lamp arm, hook together only one wire of the old and new cords *(right).* Secure the joint with electrical tape.

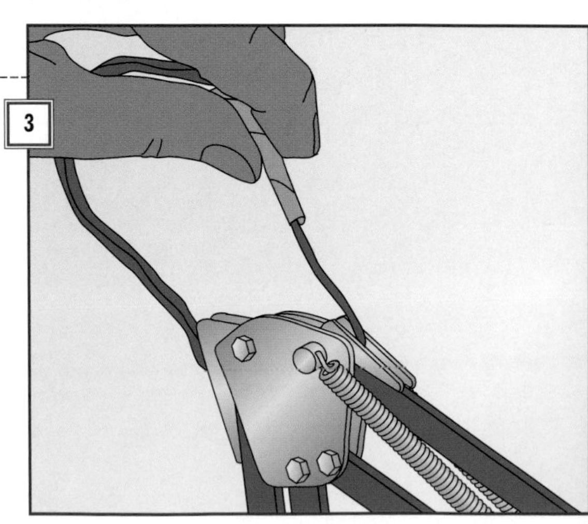


### 3. Fishing the new cord

• Feed the spliced cords into the upper lamp arm *(right).* Use the old cord to pull the new cord through the upper and lower arms.

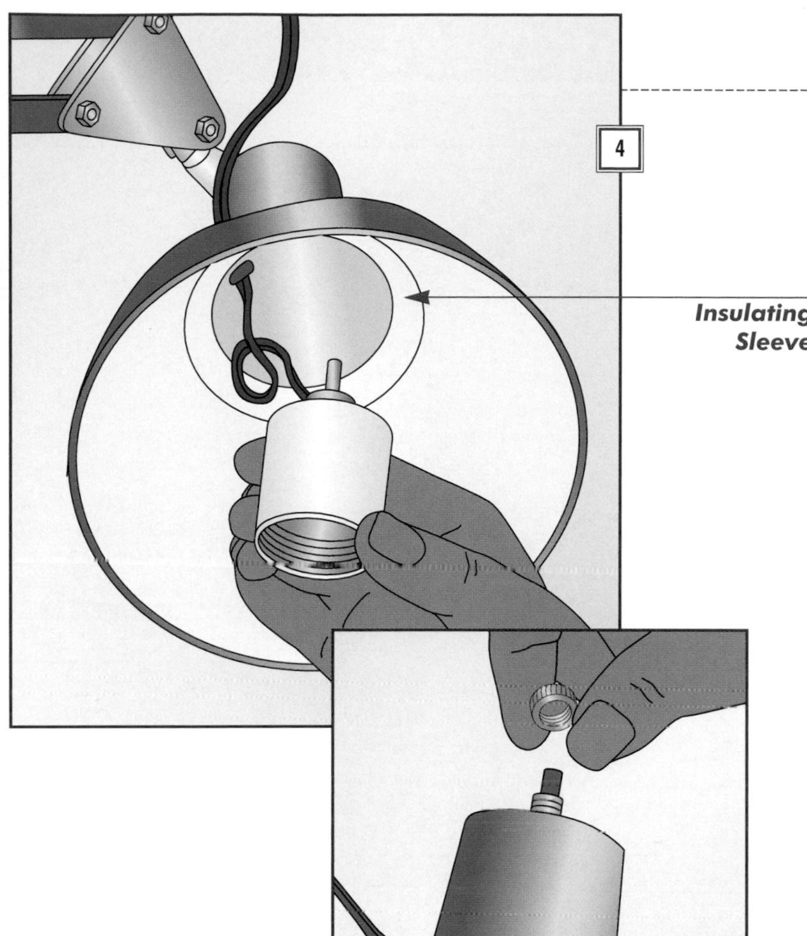


### 4. Reassembling the lamp

- Feed the end of the new cord into the lamp shade.
- Strip the wire ends, and attach them to the socket terminals *(page 21).*
- Pull the cord at the lamp elbow to create 3 or 4 inches of slack.
- Set the socket assembly inside the shade *(left).* Screw on the socket-retaining ring *(inset)* and tighten with pliers.
- Remove all but 2 inches of slack at the elbows by pulling the cord through from the base.
- Install a new plug *(pages 26–29).*

## TROUBLESHOOTING A HALOGEN LAMP

Start with the same tests you would perform on an incandescent lamp *(pages 19, 23, and 25).* Then unplug the lamp and remove the bulb with a clean cloth *(right, top).* (Touching a halogen bulb with your fingers deposits oils that shorten its life; if you accidentally touch it, clean the spot with rubbing alcohol.)

Set a multitester to the AC scale just above the lamp's rated voltage (test a low-voltage lamp on the 50-volt scale); then touch the probes to the socket terminals *(right, bottom).* If the meter shows the rated voltage, try a new bulb. The cause of lower-than-rated voltage may be a faulty transformer (on low-voltage lamps), a blown lamp fuse, a broken cord, or a damaged socket.

Models vary greatly, and replacement parts can be hard to find; so it's usually best to take halogen lamps to a dealer for repairs.

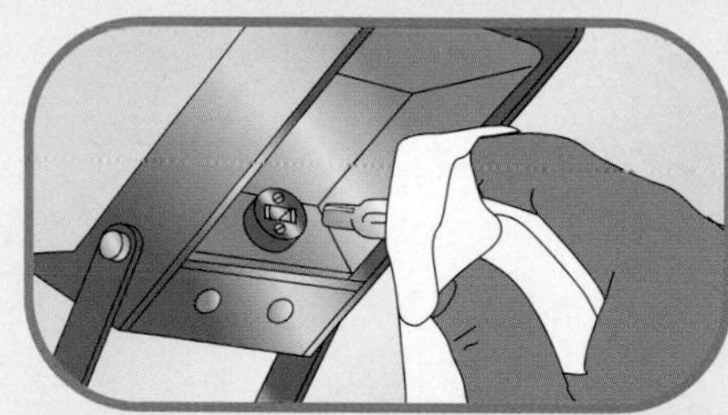

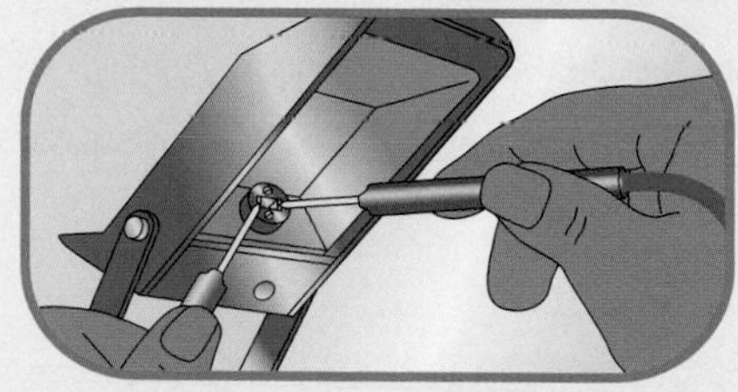

# Troubleshooting a Fluorescent Desk Lamp

### Checking the Tube and Removing the Lamp Head

• With the lamp switch in the OFF position, grasp the tube *(inset)* as close to the base as possible. Gently wiggle it back and forth until it pulls free.

• Take the old tube with you to purchase a replacement.

• Holding the new tube at the base, push it firmly into the socket.

• If the lamp still doesn't work, unplug it. Loosen any screws or clamps holding the lamp head on the stem, then slide it up and off *(right).*

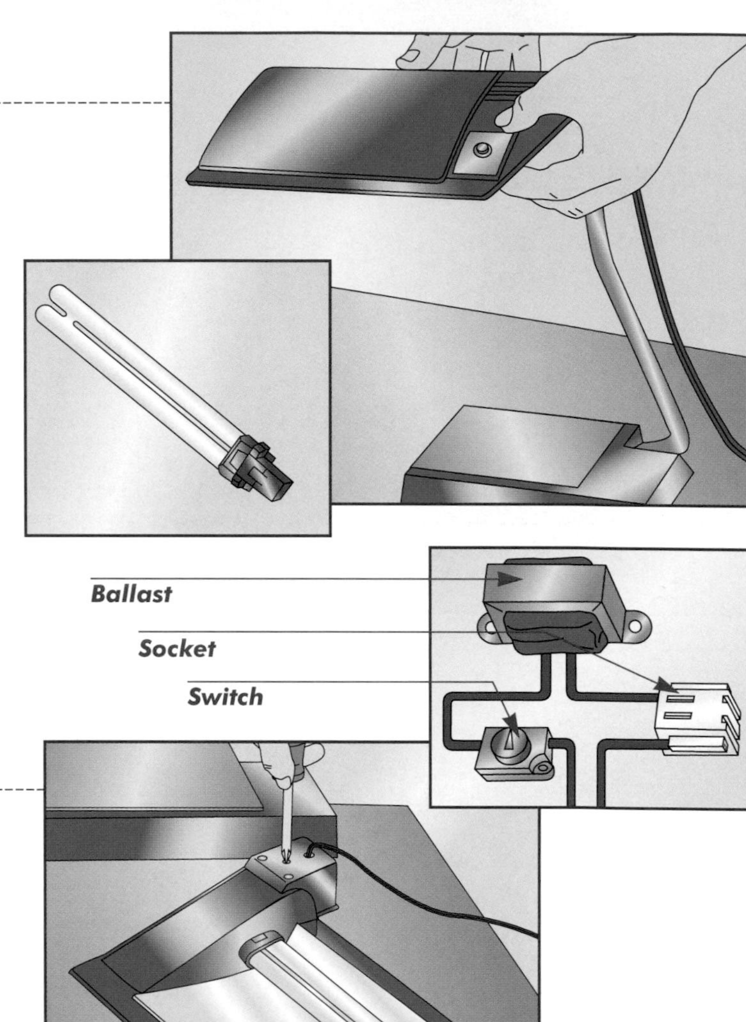


### Gaining Access to Key Parts

• Remove any screws from the component case in the lamp head *(right),* and pry off the cover with a screwdriver.

• Note how the ballast, socket, and switch are connected; also note their positions in the lamp head *(inset).*

### Testing and Replacing the Socket

• The single socket on this lamp has two push-in terminals. To disconnect the socket, push a pin or paper clip into the slot beside the wire and pull the wire free *(right).*

• The socket is secured by two plastic prongs. Squeeze them together and pull the socket out of the lamp.

• Buy a replacement socket of the same wattage, and snap it into place.

• Reattach the cover plate, and return the assembled head to the lamp arm.

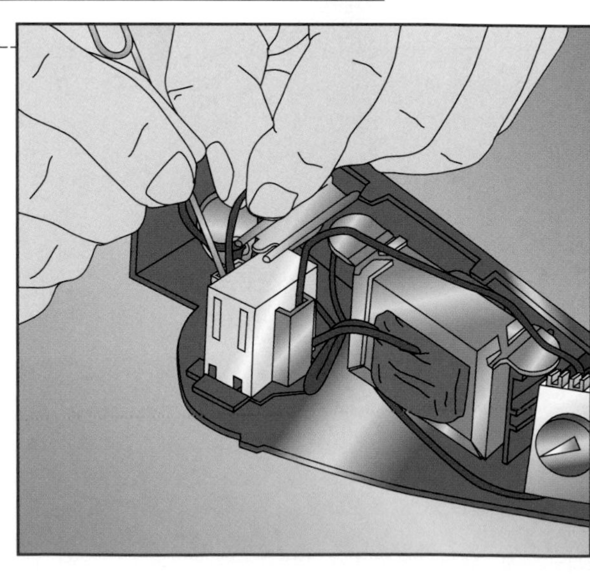

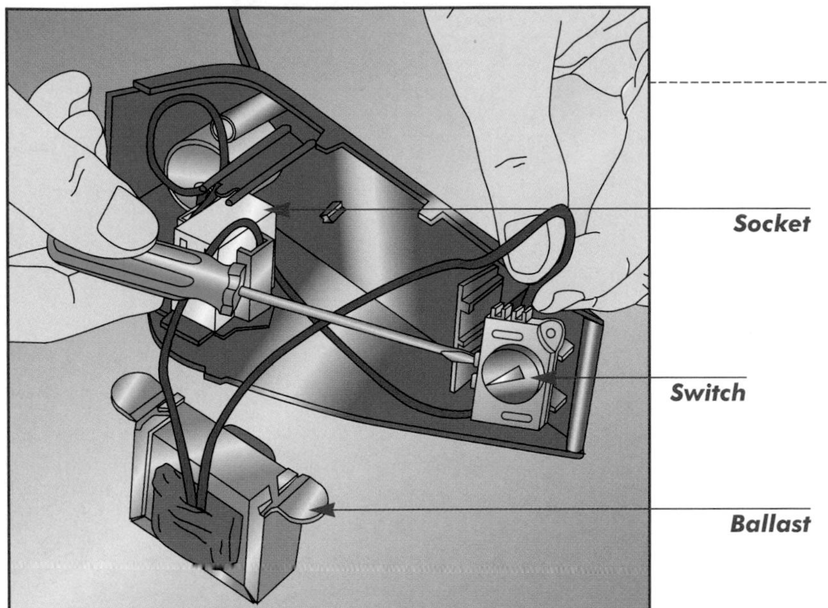


### Replacing the Switch or Ballast

- Disconnect the switch by detaching the wires from the two screw terminals *(left)*. Detach the ballast wires from the switch and socket.

- Take the old ballast or switch to a lamp store for a matching replacement.

- Connect the new components to the same wires as the old.

- Set all of the parts back into the lamp head, tuck in the wires, and reattach the cover plate.

- If the lamp still doesn't work, test the cord and plug *(page 30)*, and if either part fails a continuity test *(page 20)*, replace it.

## REPAIRING A TRIGGER-SWITCH FLUORESCENT LAMP

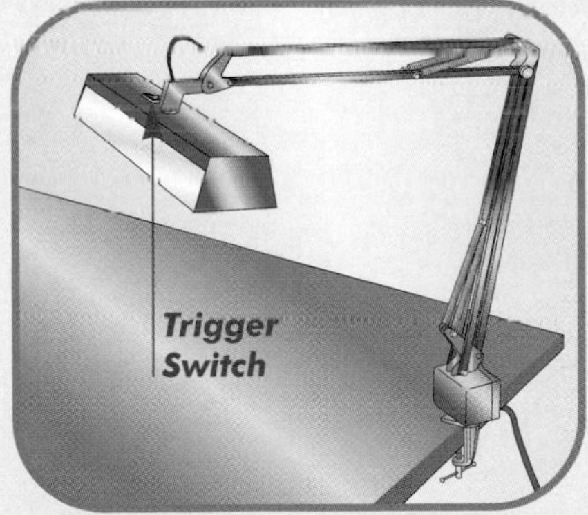


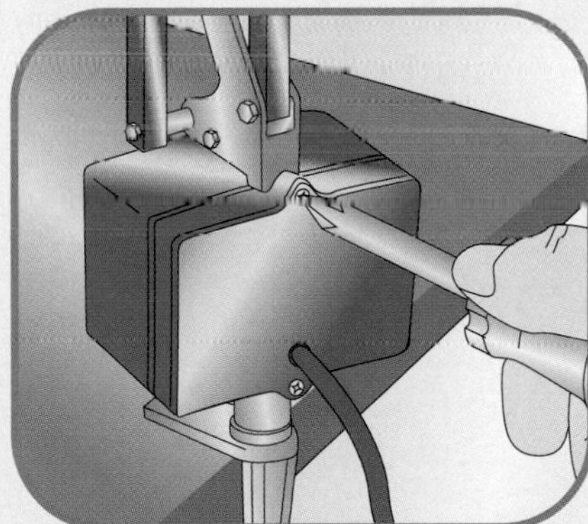

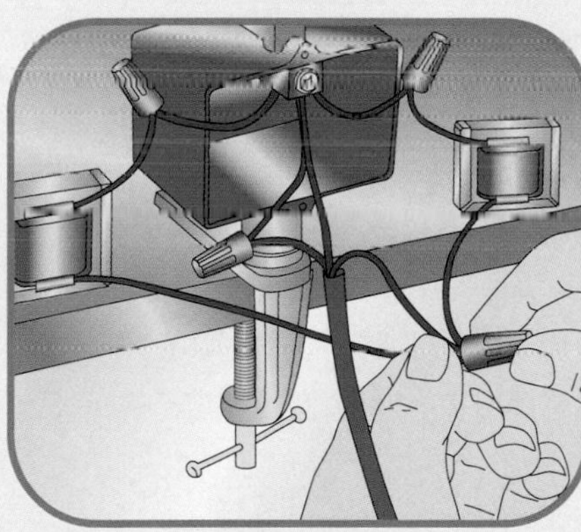

Trigger-switch fluoresent lamps, like the one shown above, left, are similar to starter-type fixtures *(page 56)*: Pressing the switch button for several seconds provides the necessary voltage surge to energize the gas in the bulb. If a trigger-switch lamp won't light, first replace the tube. If that fails, unplug the lamp and perform continuity tests on the plug, cord, socket, and switch *(page 20)*. Replace components as needed. In the event that all of these components pass the continuity tests, the culprit is probably the ballast, usually located in the lamp base.

Remove the screws from the base plate *(above, center)*, and pull the wires from the base *(above, right)*. Remove the ballast; some lamps have two of them. Detach the two wires from one ballast at a time. If the wires aren't color-coded, twist them together to identify them as a pair. Take the ballasts to a lamp store for a correct match. Reverse this process to reassemble the lamp.

# FIX IT: Lighting Fixtures

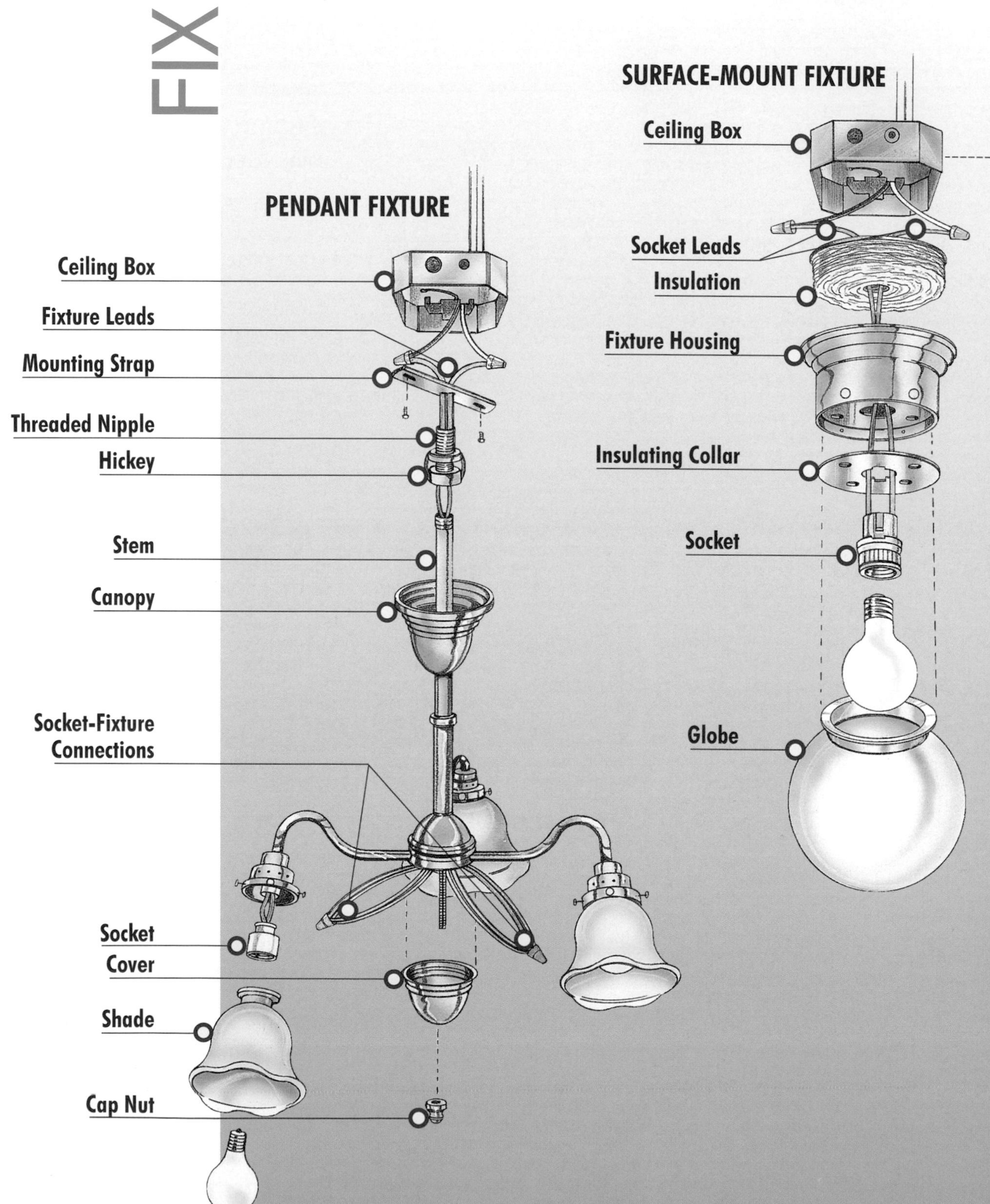

# Chapter 3

## Contents



## How They Work

**Incandescent ceiling lights come in two varieties—surface-mount fixtures *(near left)* and pendant fixtures, or chandeliers *(far left)*. Both have leads that connect to house-circuit wires in a ceiling box. The leads of a surface-mount ceiling fixture attach directly to the socket. In the pendant fixture, leads from the sockets connect to fixture leads that run up the stem.**

**A surface-mount fixture fastens directly to the ceiling box with screws through the housing. Pendant fixtures—as well as ceiling fans *(page 62)*, which often incorporate a light fixture—hang from a mounting strap fastened to the box. A threaded nipple screws into the strap and is attached to the stem with a fitting called a hickey. Other fixture-mounting variations include pull-chain sockets *(page 45)*, track lights *(page 50)*, and recessed fixtures *(page 53)*.**

**A fluorescent fixture *(page 56)* differs from the incandescent variety mainly in socket design—plus the presence of a transformer called a ballast, which raises voltage high enough to excite the gas inside a fluorescent tube to produce light.**

# Troubleshooting

| Problem | Solution |
|---|---|
| • Incandescent light fixture flickers or does not light | Tighten or replace bulb • Replace fuse or reset circuit breaker **10** • Check socket tab **43** • Test and replace socket **43–46, 48–49, 53–55** Test socket switch, replace fixture (pull-chain sockets) **45–46** • Test and replace socket and stem wires (pendant fixtures) **48–49** • Test wall switch **68** • |
| • Track-light fixtures flicker or do not light | Tighten or replace bulb • Replace fuse or reset circuit breaker (if no fixture lights up) **10** • Check socket tab **43** • Clean track contacts **50** • Test and replace socket **51–52** • Check wall switch (if no fixture lights up) **68** • |
| • Fluorescent fixture does not light | Replace fuse or reset circuit breaker **10** • Tighten or replace tube **57–58** • Replace starter (if applicable) **58** • Replace socket **59** • Replace ballast **60** • Replace fixture **61** • |
| • Fluorescent tube glows at center but not at ends | Tighten or replace tube **57–58** • Reverse tube in sockets (if only one end is dark) • |
| • Fluorescent fixture blinks or flickers or is slow to light | Location is too cold • Tighten or replace tube **57–58** • Replace starter (if applicable) **58** • Replace ballast **60** • |
| • Fluorescent fixture hums | Tighten ballast-mounting screws • Replace ballast **60** • |
| • Fluorescent fixture seeps black resin | Replace ballast **60** • |
| • Ceiling fan does not work | Replace fuse or reset circuit breaker **10** • Test motor **63** • Adjust and replace parts **63** • |
| • Ceiling fan light does not work | Tighten or replace bulb • Replace fuse or reset circuit breaker **10** • Check socket tab **43** • Test socket switch, replace fixture (pull-chain models) **45–46** • Test and replace socket and stem wires **48–49, 55** • Test wall switch **68** • |
| • Ceiling fan wobbles excessively | Inspect and replace cotter pin on motor housing **63** • Inspect and replace cotter pin on mounting ball **63** • Tighten screws on blade holders **63** • Inspect and replace blades **63** • |

# Before You Start

Whether a light fixture is an incandescent one or a fluorescent one, always look for the simpler problems before dismantling a fixture to look for more challenging ones.

**FIRST THINGS FIRST**

When a fixture doesn't light, first check the bulb. If you're not sure whether the bulb is good, screw it into another fixture. If the bulb lights, the problem with the fixture may be in the switch. Run the tests appropriate for switches *(Chapters 2 and 4)* before disassembling the light fixture to get at the sockets and wiring.

Most of the light fixtures shown in this chapter fasten to the ceiling, including the one attached to a ceiling fan. (This chapter also shows you how to check out the motor.) But the techniques for repairing them apply to wall fixtures as well. The only differences lie in the way you remove the fixture from the wall to reach the wiring, which varies as greatly among wall fixtures as it does among those fastened to the ceiling.

## Before You Start Tips:

- Many light fixtures are wired with lamp cord. A ridge in the insulation for one of the two wires designates it as the neutral wire, which always goes to the silver socket terminal or to a white circuit wire.

- When rewiring a light fixture, select new wires of the same gauge as the old ones, with insulation that's resistant to high temperatures.

Screwdrivers
Multitester
Multipurpose electrical tool
Lineman's pliers
Diagonal-cutting pliers
Putty knife

Lamp cord
Fine sandpaper
Wire caps
Wire coat hanger

Always cut power to a lighting circuit at the service panel as the first step in a light-fixture repair. But before doing so, turn on the switch that controls the fixture, to avoid misleading readings when testing for voltage.

# Repairing Surface-Mount Fixtures

## 1. Removing the globe

The electrical test and repair procedures for surface-mount ceiling fixtures and wall fixtures are similar, although gaining access to the socket and wiring may differ.

- Flip off the wall switch and loosen any retaining screws holding the globe in place *(right),* then remove the globe.

- Tighten a loose bulb or replace one that's burned out with a new one, and flip on the switch. Reinstall the globe if the bulb lights. If it doesn't, dismount the fixture as shown in Step 2.

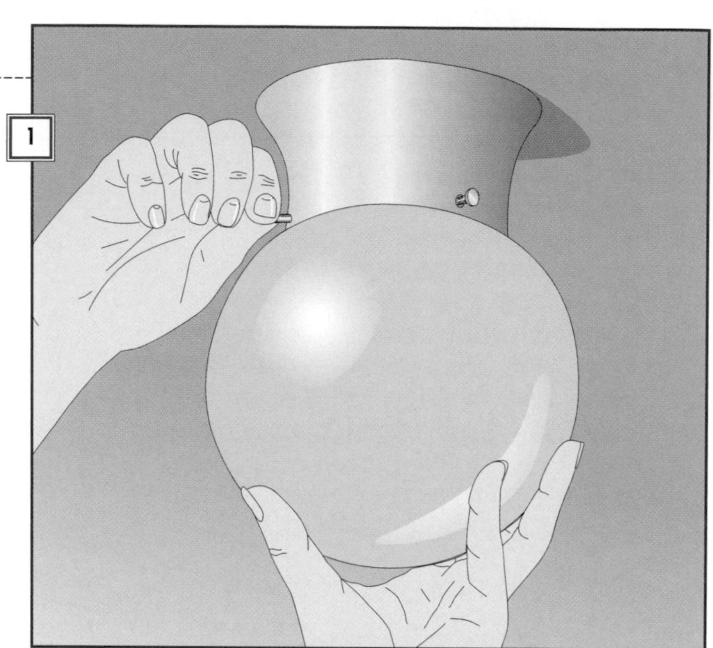


## 2. Dismounting the fixture

- Flip on the switch, then turn off power to the fixture at the service panel by setting the circuit breaker to the OFF position or removing the fuse *(page 10).*

- Remove the mounting screws from the base, and gently lower the fixture *(right).*

- For a wall fixture *(inset),* remove the globe for access to the locknut. Unscrew the locknut, then gently pull the fixture housing off the threaded nipple.

- Take out the screws that hold the mounting strap in place to gain access to the wiring connections.

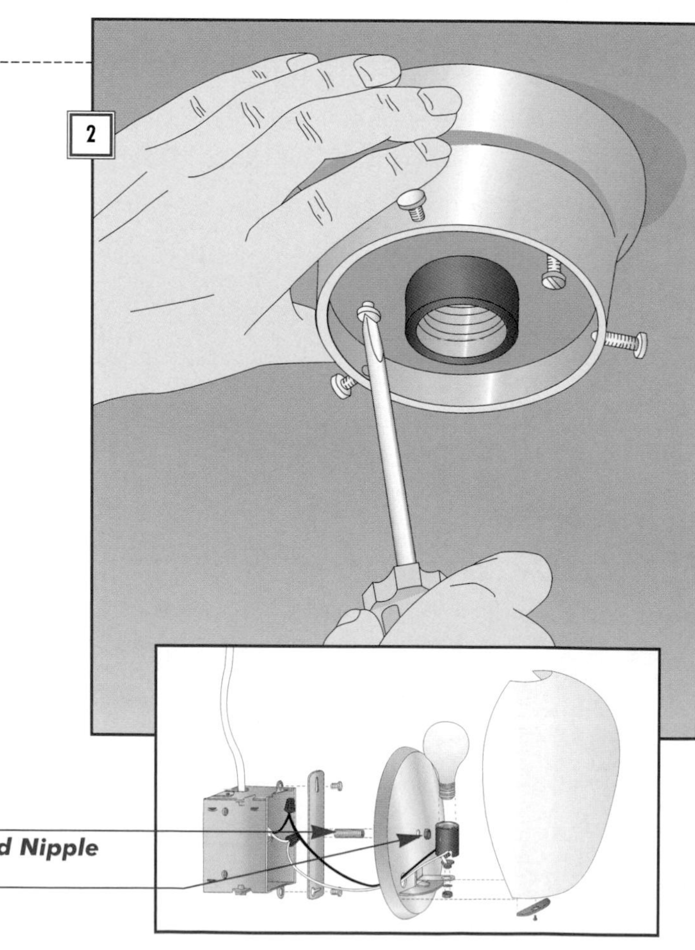

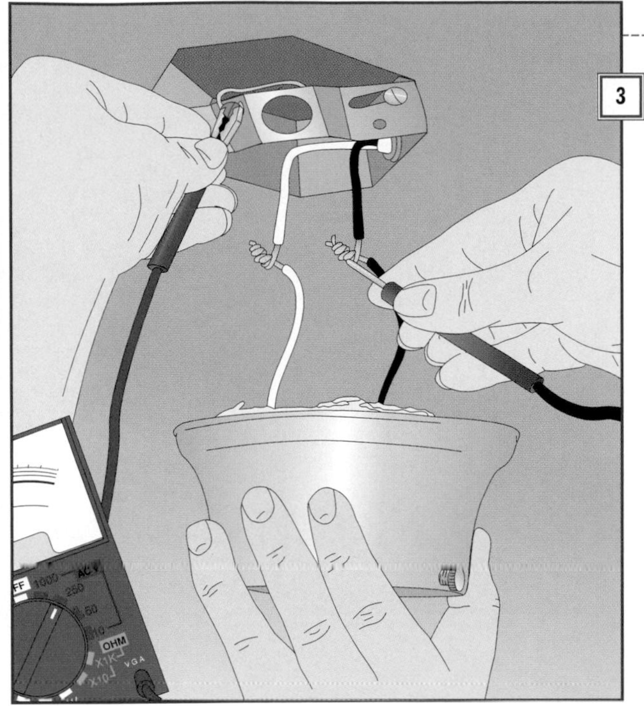


## 3. Checking for voltage

• Have a helper support the fixture while you unscrew the wire caps from both the black and white wire connections, being careful not to touch the bare wire ends.

• Set a multitester to 250 in the AC-voltage range *(page 20).* Touch one probe to the ground screw on the mounting strap and the other first to the black wire connection *(left),* then to the white wire connection. Next, touch probes to the white wire and black wire connections. If the needle moves in any test, return to the service panel and turn off the correct circuit.

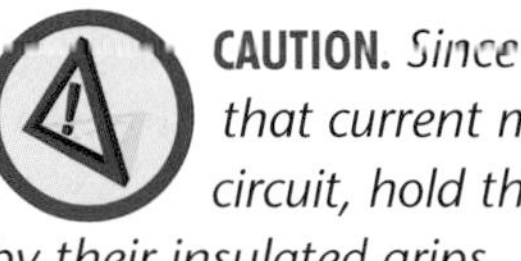

**CAUTION.** *Since there is a possibility that current may be flowing to the circuit, hold the multitester probes by their insulated grips.*

• After confirming that power is off, untwist the connections and take down the fixture.

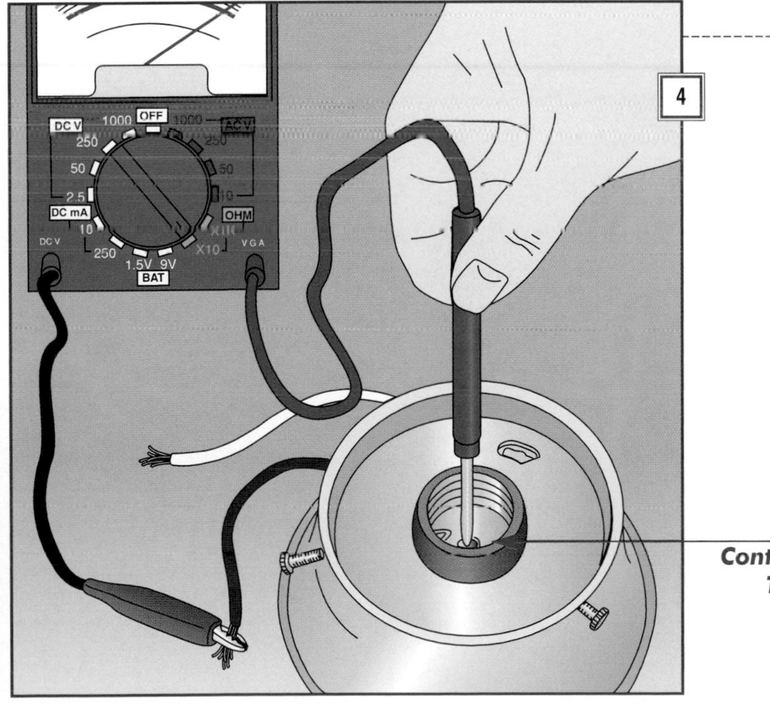


## 4. Testing the socket

• Expose the base of the socket by pulling away any insulation. Scrape any corrosion from the socket contact tab and pry it up slightly to make better contact with the bulb *(page 18).*

• Set a multitester to RX1 and test for continuity *(page 20).* To do so, clip one probe to the bare end of the black wire and touch the second probe to the contact tab *(left).* A reading of 0 ohms indicates continuity between the wire and the contact tab.

• Clip a probe to the bare end of the white wire, and touch the probe to the socket's threaded metal tube. The meter should indicate continuity.

• Replace the socket and its leads if you obtain different results.

## 5. Replacing the socket

• A spring clip secures some sockets. Press it with a screwdriver to free the socket, then pull the socket and wires out of the housing. If the socket is the two-part type with a knurled retaining ring *(page 45)*, unscrew the retaining ring and remove the socket.

• Buy a compatible replacement socket, and thread the socket wires through the insulating collar (if any), then through the fixture housing *(right)*.

• Push a new spring-clip socket into the fixture; stop when you hear the clips snap into place. Attach a two-part socket by fitting the bottom half through the back of the fixture and screwing the top half onto it from the front.

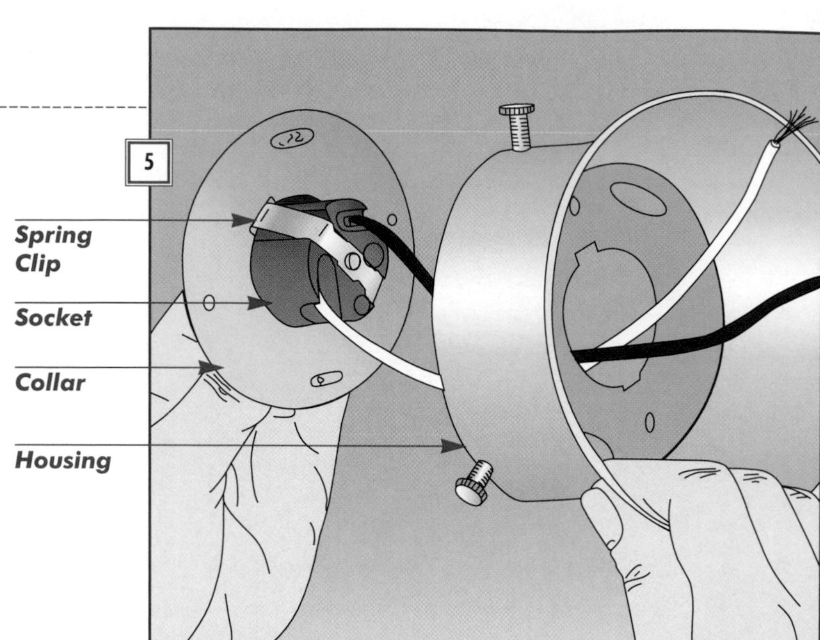


## 6. Connecting the wires

• Suspend the fixture from the electrical box by a hook fashioned from a wire coat hanger, and restore any insulation that you removed *(right)*.

• Twist together the black socket lead with the black wire from the ceiling box, and the white socket lead with the white wire from the box; screw a wire cap onto each connection *(page 51)*. Make sure the cable's ground wire is secured to the ground terminal on the mounting strap.

• Remove the coat hanger, and fold the wires into the ceiling box.

• Align the mounting slots with the mounting holes on the ceiling box, then insert and tighten the mounting screws. Screw in the bulb, set the globe in place, and retighten the retaining screws.

• Restore power to the circuit.

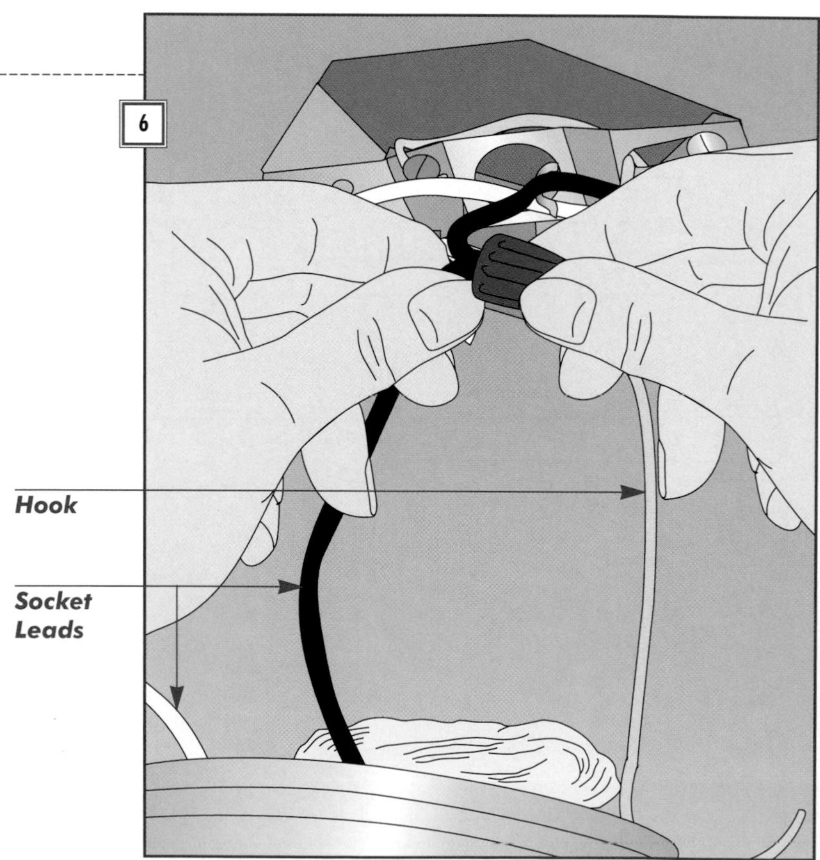

# Replacing a Pull-Chain Socket

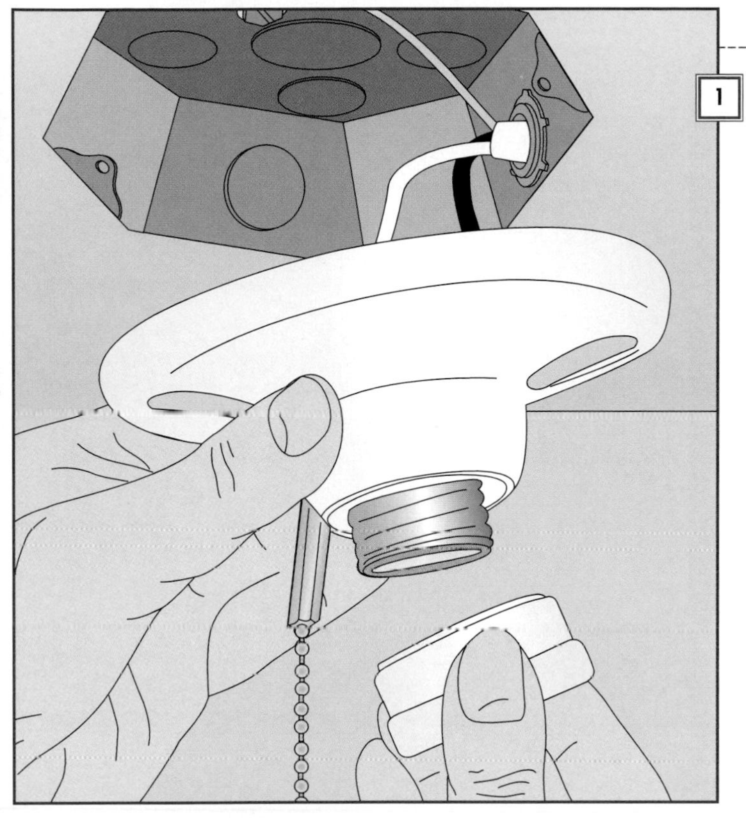

## 1. Gaining access to the socket

- Turn off power to the fixture by unscrewing the fuse or tripping the circuit breaker *(page 10)*. Remove the bulb from the socket.

- Take out the screws that hold the fixture to the ceiling, and unscrew the cap from the socket *(left)*. Lower the fixture, taking care not to touch the socket or any metal parts until you have confirmed that the power is off *(Step 2)*.

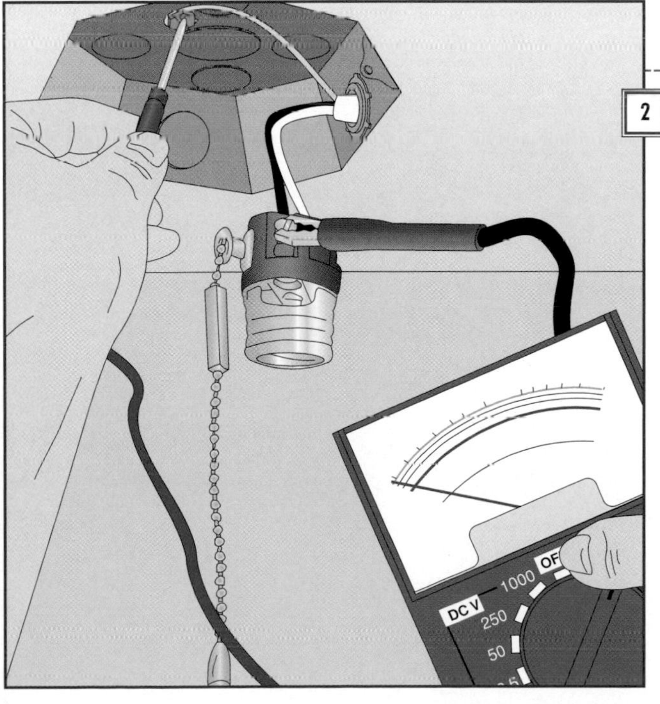


## 2. Testing for voltage

- Set a multitester to 250 in the AC-voltage range, and test for voltage *(page 20)*. Hold the probes by their plastic sheathing to avoid the possibility of electric shock.

- Clip one probe to the brass socket terminal, and touch the other to the ground wire. Repeat the test between the silver terminal and the ground wire *(left)*. If the needle moves in either test, return to the service panel and cut power to the correct circuit.

- After confirming that the power is off, disconnect the wires from the socket terminal screws.

## 3. Testing the Socket and Switch

- Set a multitester to RX1 for a continuity test *(page 20).* Clip one probe to the threaded metal tube, and the second probe to the silver terminal *(right).* The meter should indicate continuity.

- Clip one probe to the brass terminal screw, and the second probe to the socket contact tab.

- Pull the switch. The meter should indicate continuity when the switch is in one position and no continuity in the other. Get a new socket if your results differ.

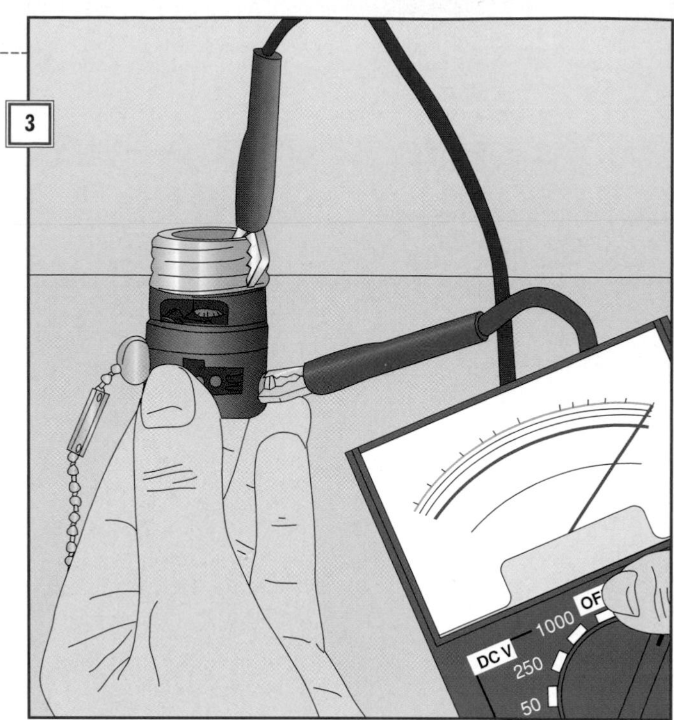


## 4. Replacing the Socket

- Whether installing a new socket or an old one, brighten dirty wires with sandpaper. Trim and restrip burned wire ends, then bend them into hooks to make a solid contact with the terminals *(page 69).*

- Connect the black wire to the brass socket terminal, and the white wire to the silver one. Carefully fold the wires into the ceiling box, and thread the pull chain through its hole in the fixture *(right).*

- Align the mounting slots with the mounting holes on the ceiling box, then insert and tighten the mounting screws.

- Screw in a bulb, and restore power to the circuit.

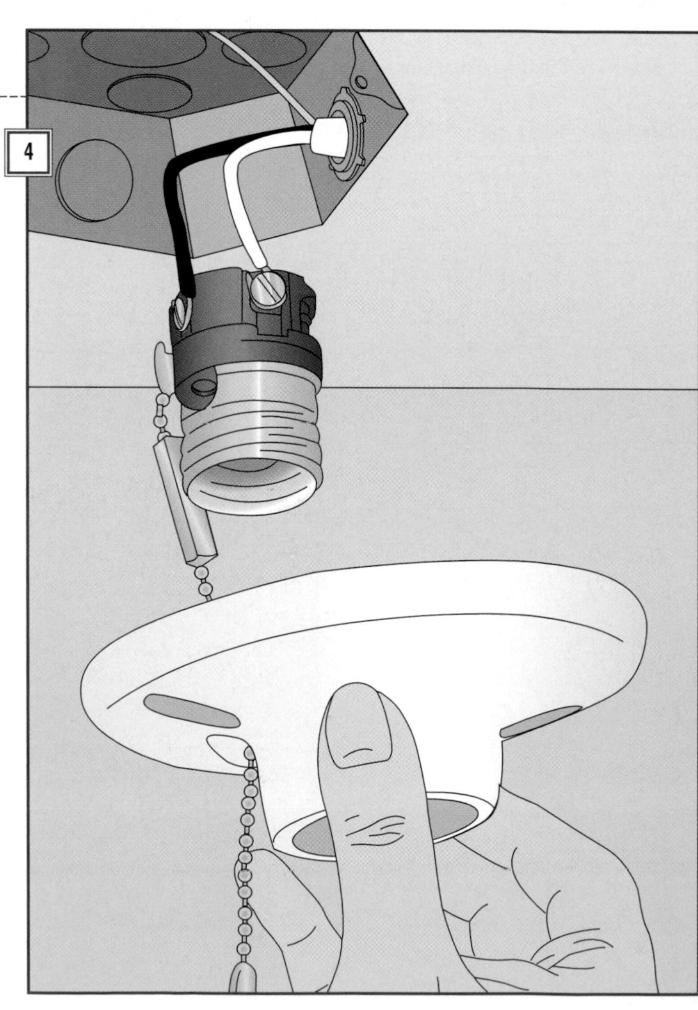

## CHOOSING THE RIGHT LIGHTBULB

Indoor incandescent lamps and fixtures use a standard, A-type bulb with a threaded base that screws into a medium-base socket. The 3-way bulb, similar in appearance to the A-type bulb, requires a special socket. Two filaments inside this type of bulb light one at a time for low- and medium-intensity lighting, or together for maximum brightness. Track and recessed fixtures usually require reflector bulbs, which provide an intense, concentrated beam.

For the demands of outdoor lighting, there are special bulbs made of shatterproof glass designed to withstand sudden temperature changes. Most outdoor lamps and fixtures come with medium-base, water-resistant sockets and take A-type bulbs. Other types of outdoor bulbs include floodlights; Christmas lights, which fit into small, threaded candelabra sockets; and low-voltage bulbs that fit threaded or bayonet sockets.

Various lightbulbs are sold in clear, frosted, and coated versions, providing different qualities of light. Bulbs are also available in different wattages; the higher wattages produce more light but use more electricity. For stairways, hallways, outdoor fixtures, and other hard-to-reach places, consider buying long-life bulbs. The filaments of these bulbs burn at a lower temperature, increasing the life of the bulb to up to three times that of a regular bulb. Another option is the bulb-life extender, a small button-like disc that can add years to the life of a standard A-type bulb by limiting the surge of electricity to the bulb when you turn it on. To install a bulb-life extender, simply peel off the backing and stick it to the base of the bulb before threading the bulb into the socket.

When buying a replacement bulb, consider your lighting requirements, then check the size of the socket and its recommended maximum wattage. The wattage information is usually found on a sticker near the socket of most lamps and fixtures. Never use a bulb that exceeds the socket's maximum recommended wattage; if you do so, the heat produced could melt a plastic globe or shade or could damage wire insulation. Because air space in a globe is essential, avoid using a bulb that is too large for the lamp or fixture.

A-type

The most common type of light bulb. Used in standard lamps and fixtures. Long-life and weather-resistant outdoor versions are available. For medium-base socket; 4–300 watts.

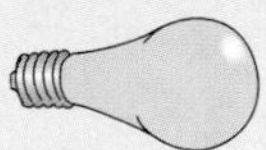
3-way

For lamps with 3-way sockets; provides low, medium, and high light intensities. For medium-base sockets; 30/70/100, 40/75/150, and 100/200/300 watts.

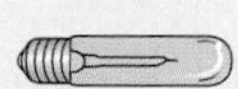
T (tubular)

Found in medium-intensity desk lamps; in showcases, mounted above a picture frame; and in aquarium canopies. For medium-base socket; 15–150 watts.

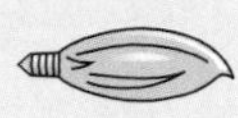
Candelabra

For decorative lamps and fixtures where low light or accent lighting is desired. For medium-base and candelabra socket; 15–60 watts.

Globe

A decorative bulb that does not require a shade. Used around makeup mirrors and in hanging fixtures. For medium-base and candelabra socket; 40–150 watts.

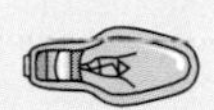
Night-light

Small, low-wattage bulb used in plug-in fixtures to illuminate hallways at night. For medium-base and candelabra socket; 4,7,7-1/2 watts.

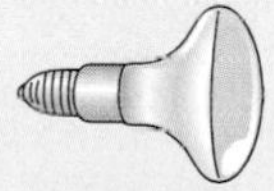
R (reflector)

Also called spotlighst. A built-in reflector directs the beam. Used where directional lighting is required (such as track lighting). For medium-base socket; 15–60 watts.

Floodlight

Used in track lighting, recessed fixtures, and outdoors. For medium base socket; 25–250 watts. Halogen versions are also available.

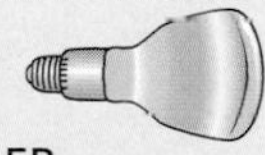
ER

Found in recessed, downlight fixtures. For medium-base socket; 50–120 watts.

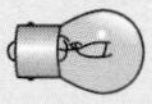

Low-voltage

Common in indoor fixtures and outdoor fixtures. Requires a transformer to reduce 120-volt household current. For medium-base, candelabra, and bayonet socket; 6–16 volts.

# Testing and Repairing a Pendant Fixture

## 1. Uncovering the Ceiling Box

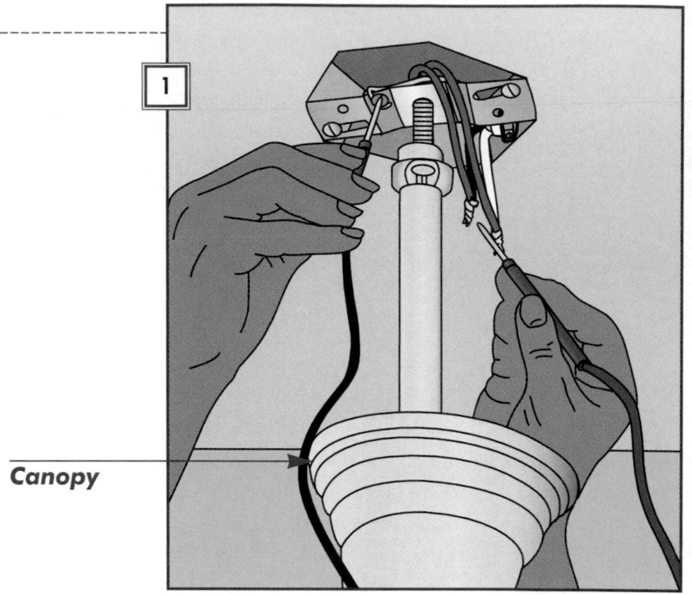


• Flip on the switch, and turn off power to the circuit at the service panel *(page 10).* Loosen the canopy screws and lower the canopy.

• Unscrew the wire caps from the connection, taking care not to touch the wires.

• Set a multitester to 250 in the AC-voltage range *(page 20).* Holding the probes by their plastic handles, touch one probe to the black wire connection and the other first to the ground terminal on the mounting strap, and then to the white wire connection *(right).* Test between the white wire and ground. If the needle moves in any test, return to the service panel and turn off power to the correct circuit.

## 2. Gaining Access to the Sockets

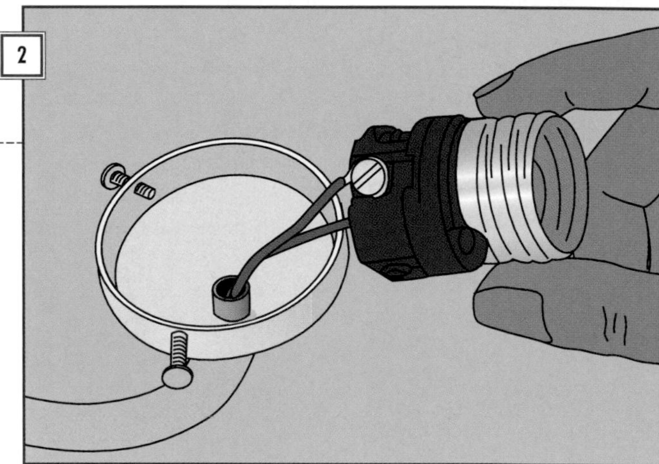


• Unscrew the socket from the fixture. Locate the word PRESS on the socket, and apply pressure to remove the outer shell.

• Slip off the socket insulating sleeve *(page 18).*

• Pull the socket away from the cap to expose at least 1 inch of wire *(right).*

## 3. Testing the Socket

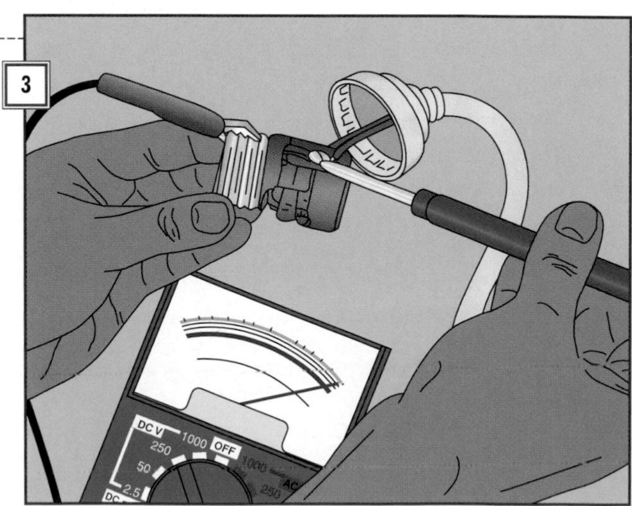


• Set a multitester to RX1, and scrape any corrosion from the socket contact tab.

• Clip one probe to the socket's threaded, metal tube, and touch the other to the silver terminal screw *(right).* Next clip one probe on the brass terminal screw and touch the socket contact tab. There should be continuity in both tests *(page 20).* If so, go to Step 4. If not, replace the socket as shown on page 46.

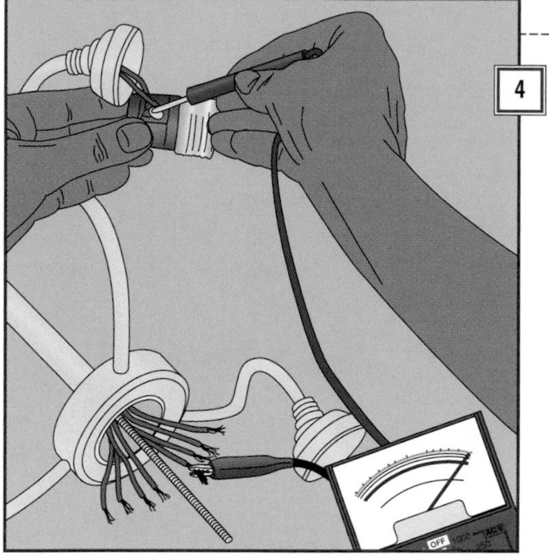


## 4. Testing socket wires

• Remove the cover at the base of the fixture. Pull out and disconnect the wires. If none of the sockets work, go to Step 5.

• To test wires to a single socket, first confirm that the ridged wire goes to the silver terminal. If not, reverse the connections.

• Set a multitester to RX1, and clip one probe on the ridged socket wire at the fixture base *(left).* Touch the other probe to the silver terminal. Next, place the alligator clip on the unridged wire and touch the probe to the brass terminal. Both tests should indicate continuity *(page 20).*

• Replace the socket wires if they fail the test. Disconnect the socket, and fish new wires to the socket *(page 30).*

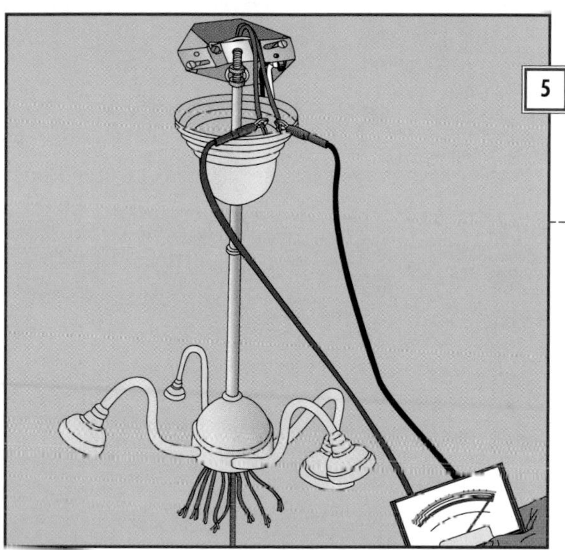


## 5. Testing the stem wires

• Twist the two stem wires together at the fixture base. Clip one multitester probe to each of the wires at the top of the stem *(left).* The meter should indicate continuity.

• If the wires fail the test, fish new wires through the stem *(page 30).*

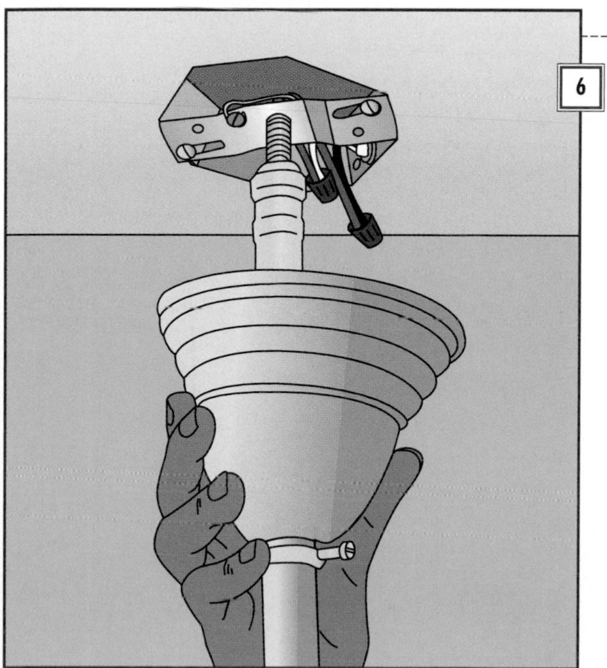


## 6. Completing the repair

• Twist the ridged socket wires together with the ridged stem wire; twist the unridged socket wires together with the unridged stem wire. Then fold the connections into the wiring compartment and screw on the cover.

• Twist the ridged stem wire together with the white wire in the box, and the unridged wire with the black wire. Secure the connections with wire caps.

• Fold the wires into the box, and slide the canopy up the stem and against the ceiling *(left).* Tighten any canopy screws.

• Screw in the bulbs and restore power to the circuit.

# Troubleshooting Track Lights

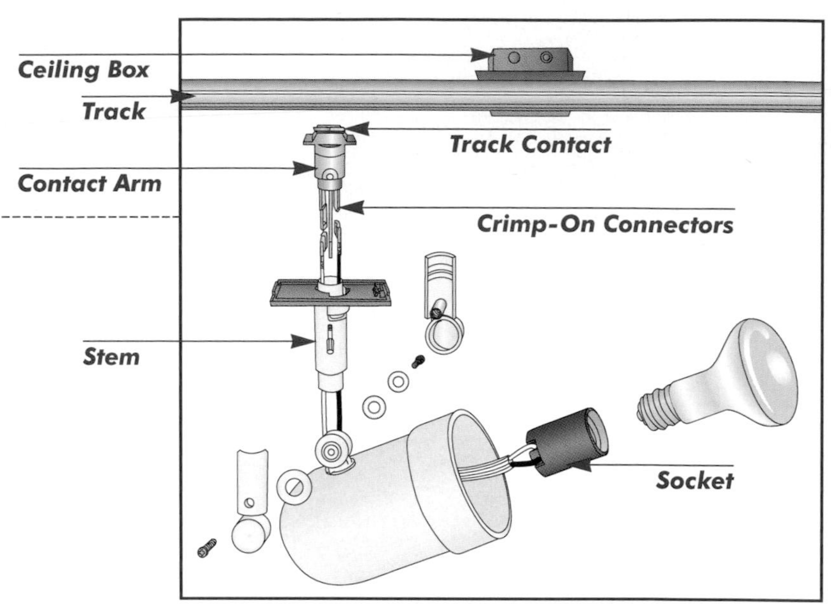


## Track-light anatomy

A track is wired to a house circuit at a ceiling box *(right),* which serves as the power source for two metal conductors that run the length of the track. Each fixture has a contact arm with two track contacts at the top. When a fixture is mounted to the track, the two contacts bridge the conductors to complete a circuit through a pair of wires to the socket and lightbulb.

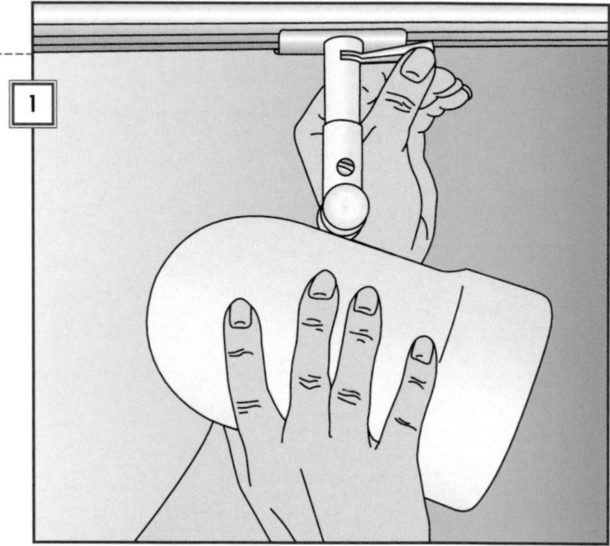


## 1. Cleaning the contacts

- Flip on the switch, and turn off power to the circuit at the service panel *(page 10).*
- Turn the lever to unlock the fixture, then turn it 90 degrees to release it *(right).*
- Clean any corrosion from the track conductors with fine sandpaper. Remove the bulb, and scrape any corrosion from the socket contact tab. Pry it up slightly.
- Screw in a bulb, reinstall the fixture in the track, and then restore power. If the bulb does not light, turn off the power, dismount the fixture, and go to Step 2.

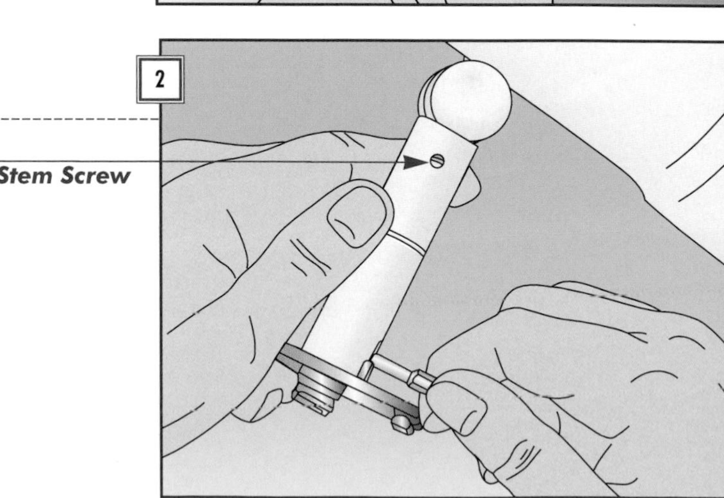


## 2. Gaining access to the socket

- Unscrew the lever by hand *(right),* and use a screwdriver to take out the stem screw.
- Remove the screws holding the socket to the shade, and pull the socket free of its mounting.

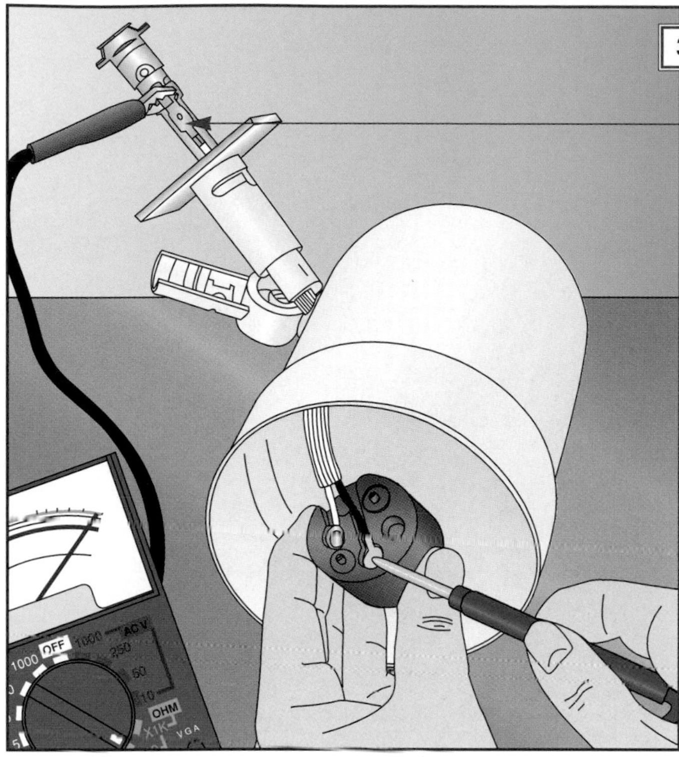


### 3. Testing the socket

Crimp-on connectors join the socket leads to contact pins protruding from the contact arm.

- Set a multitester to RX1, and clip one probe to the brass contact pin. Touch the other probe to the black wire connection at the socket *(left).* The meter should indicate continuity *(page 20).*

- Place the alligator clip on the silver contact pin, and touch the tester probe to the white wire connection. The meter should indicate continuity.

- If the socket fails either test, replace it as shown on page 52.

## Ron's TRADE SECRETS

**Splicing lamp cord to solid wire**
Here's what I do to get a good, permanent connection between stranded lamp cord and solid-copper house wires. First I strip about an inch of insulation from each wire. Then I wrap the stranded wire around the solid one and hold the two wires together near the ends with lineman's pliers *(right, top).* Next I use the pliers to fold the splice in two and squeeze it flat *(right, bottom).* Finally, I twist on a wire cap of the appropriate size, leaving no copper visible. This technique provides a solid connection and won't let the stranded wire slip.

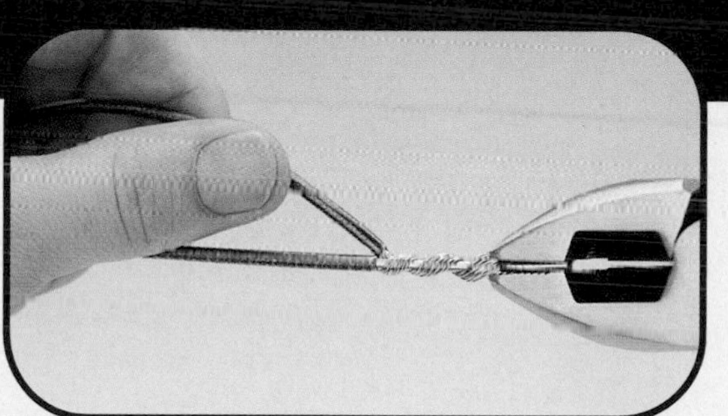

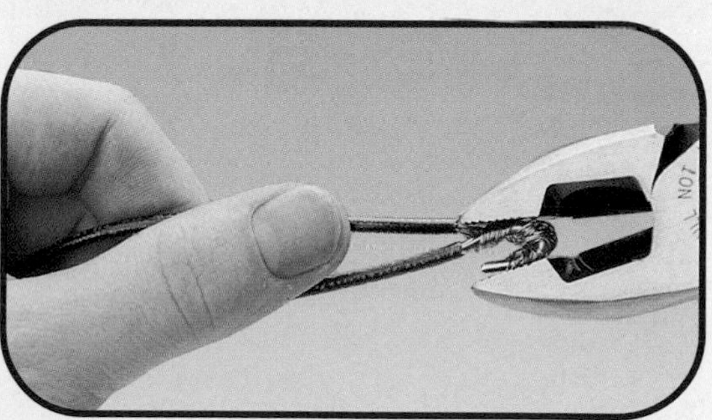

## 4. Removing the old socket

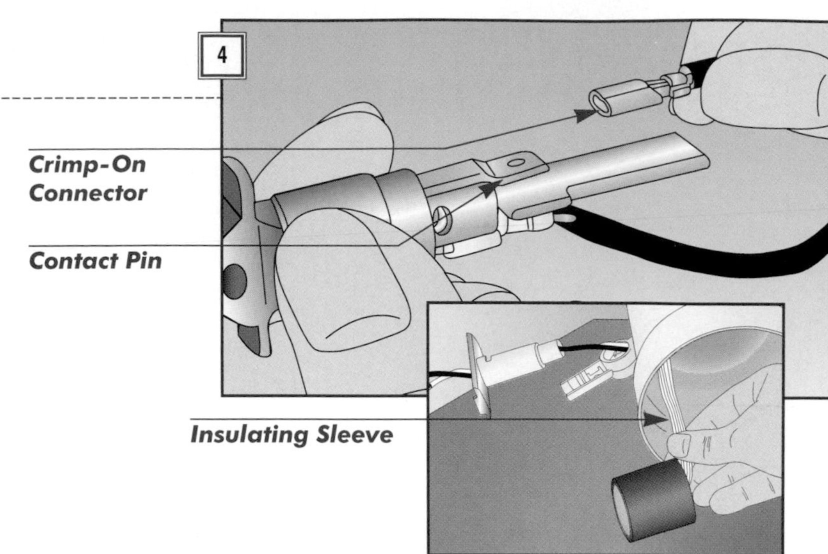


- Gently pull the socket wires from the contact pins *(right).*

- Extract the socket, wires, and insulating sleeve from the shade *(inset).* Slip the wires out of the insulating sleeve, and set it aside.

- Purchase a comparable replacement socket with preattached wires. Also buy crimp-on connectors the same size and shape as the old ones.

## 5. Installing the new socket

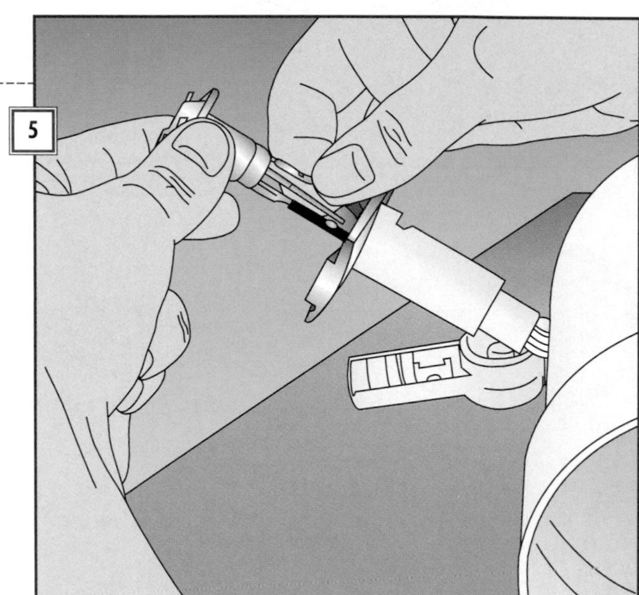


- Slip the insulating sleeve over the new wires, and thread the wires first through the hole in the shade, then through both parts of the stem.

- Twist the strands of the socket wires tightly together, and push a crimp-on connector onto the end of each. Use a multipurpose tool to crimp the connectors tightly onto the socket wires.

- Slide the black wire connector onto the brass contact pin and the white wire connector onto the silver contact pin *(right).*

## 6. Reassembling the fixture

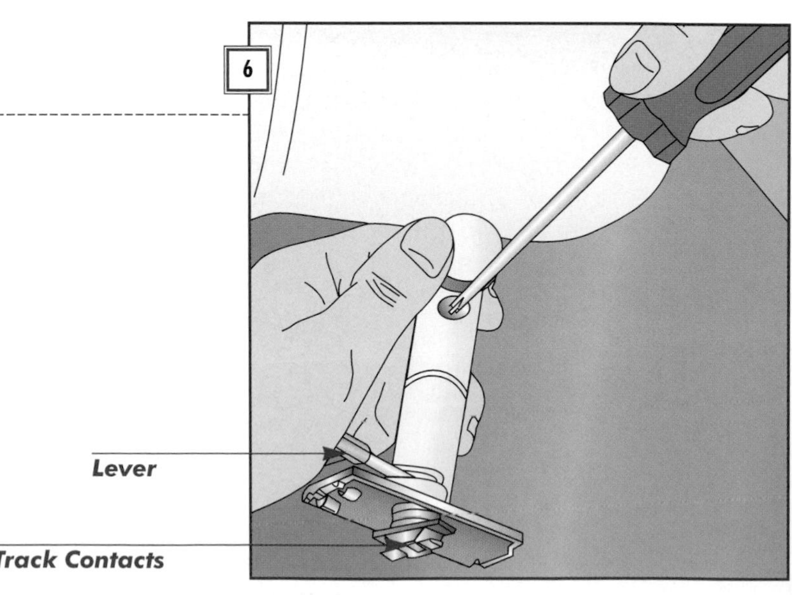


- Set the socket in the shade, and tighten the screws that secure it.

- Screw the lever into the stem by hand, then reassemble the stem and screw it together *(right).*

- Fit the fixture into the track and turn it a quarter-turn. Lock the fixture in place with the lever.

- Screw in the bulb, and restore power to the circuit.

# Repairing Recessed Fixtures

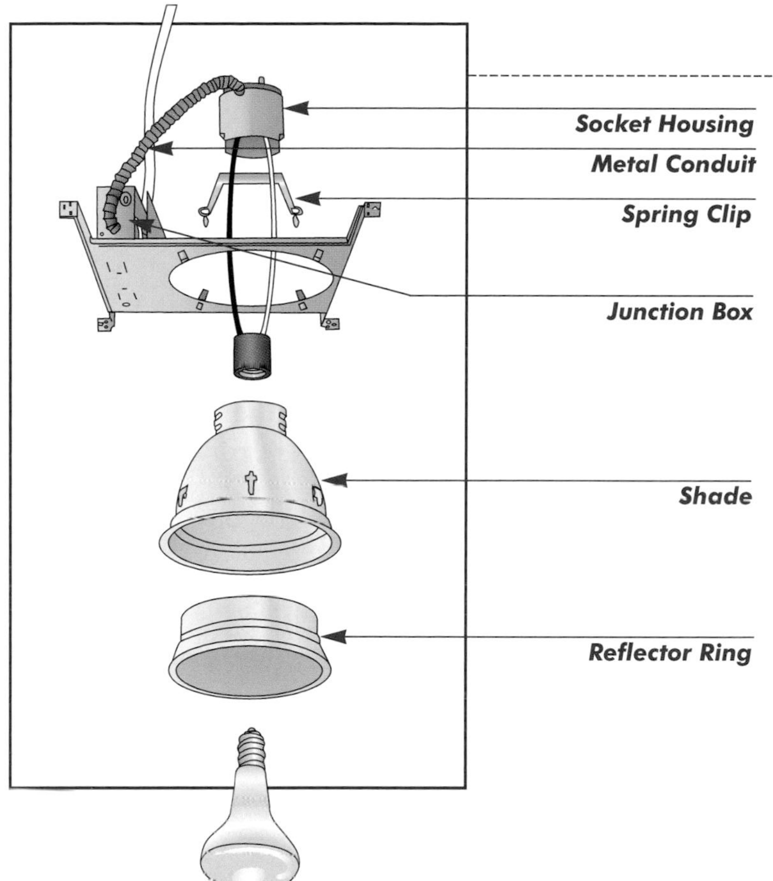


## Recessed-Fixture Anatomy

A recessed ceiling fixture is built into a housing that is installed above the ceiling. Each housing has a junction box for the connections between the house-circuit wires and the socket leads. Flexible metal conduit runs to the socket housing and protects the socket leads. A spring clip holds together the socket housing, fixture housing, and shade. A reflector ring snaps into the shade.

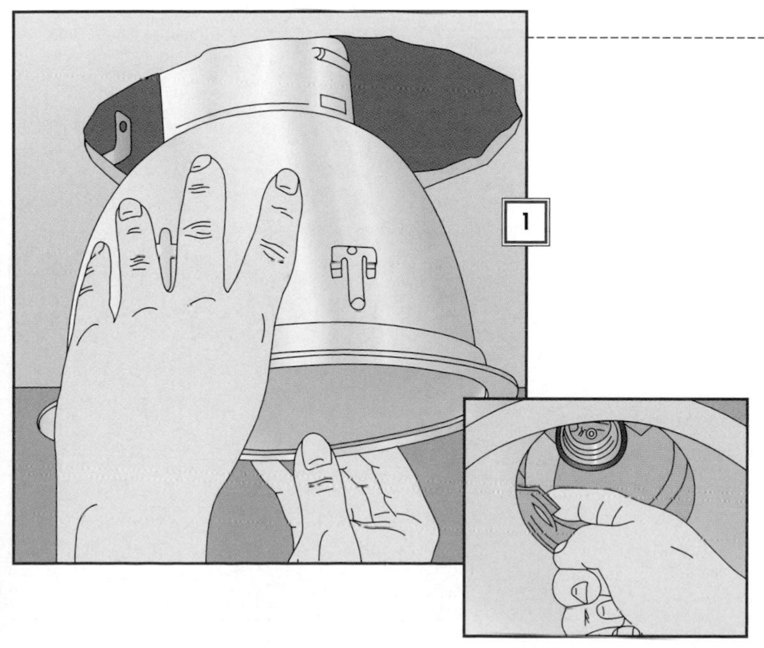


## 1. Removing the Shade

• Flip on the switch and cut power to the circuit *(page 10)*. Unscrew the bulb, then snap out the reflector ring.

• Gently pry around the shade with a stiff putty knife to work it loose, taking care not to damage the ceiling finish. Coax it down farther with your hands *(left)*.

• Push hard on the spring clips that hold the shade to the socket housing *(inset)*. Lower the shade to expose the socket housing and ceiling box.

**CAUTION:** *Do not touch the socket until you have confirmed with a voltage test that the power is off.*

## 2. Testing for voltage

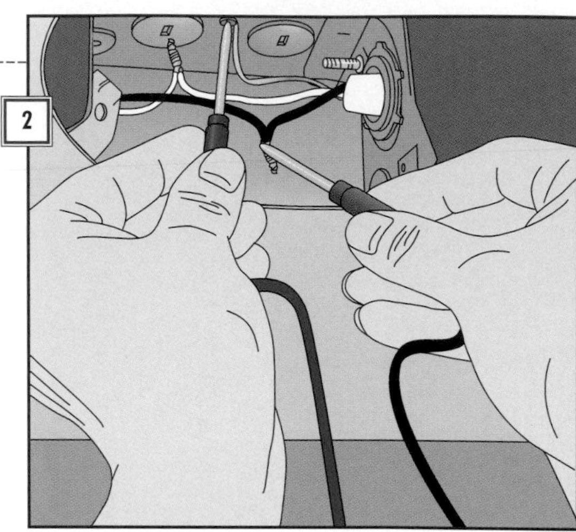


- Unclip the cover from the junction box and twist off the wire caps, taking care not to touch any bare wire ends.
- Set a multitester to 250 in the AC-voltage range *(page 20).* Touch one probe to the black wire connection, and the other first to the ground wire and then to the white wire connection. Test between the white wire connection and the ground. If the meter registers voltage, return to the service panel and cut power to the correct circuit.

## 3. Testing the socket and leads

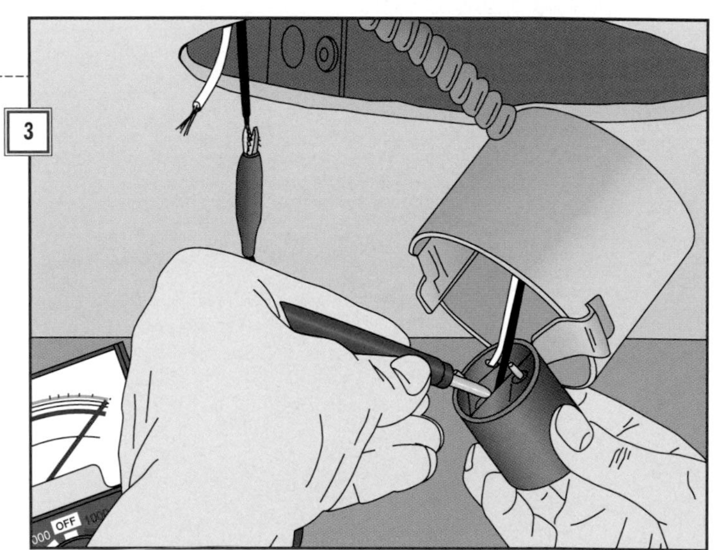


- Untwist the wire connections in the junction box.
- Set a multitester to RX1, and clip one probe to the black socket lead. Touch the other to the socket contact tab *(right).* Repeat this test between the white socket wire and the threaded metal socket tube. There should be continuity in both tests *(page 20).*
- Replace the socket *(Step 4)* if either test indicates lack of continuity.

## INSULATION AND RECESSED FIXTURES

Heat buildup around recessed fixtures can cause serious problems, especially when insulation is present. Not only does it result in premature failure of electrical parts, it sometimes leads to fires.

When you're working on recessed lights, check for insulation around the fixture. If you find any, make sure that there is at least 3 inches of air space between the insulation and the shade and junction box *(right).* If not, remove insulation to establish the necessary clearance.

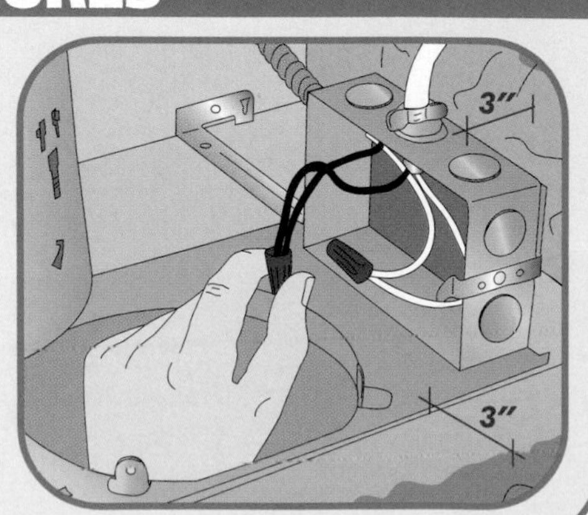

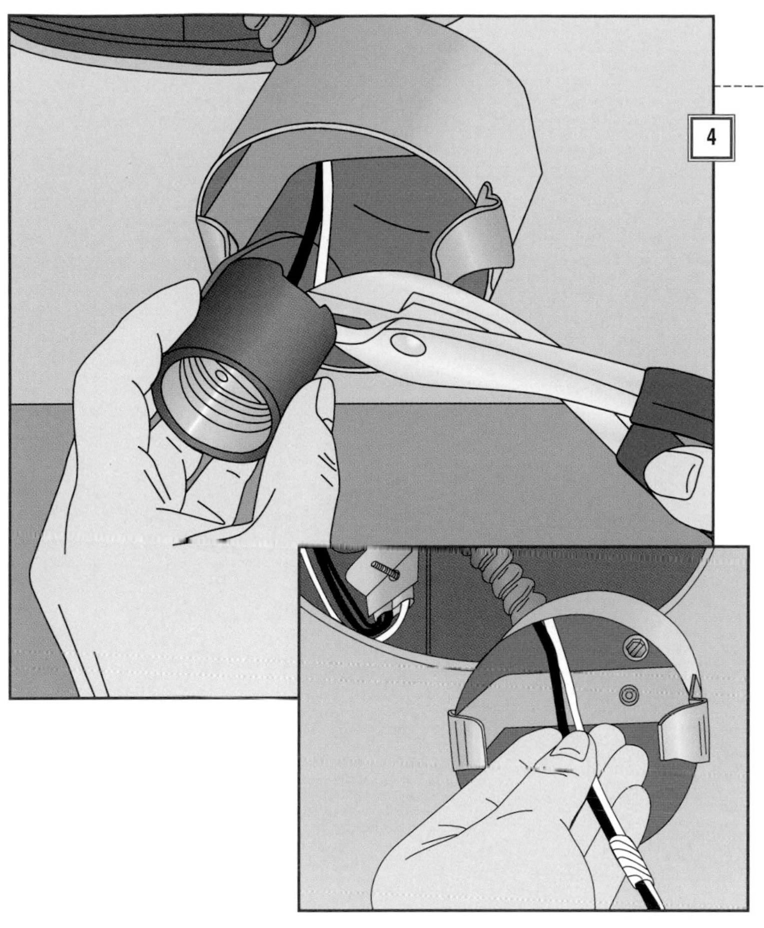



## 4. Replacing the socket

- With diagonal-cutting pliers, cut the socket leads *(left).*
- Temporarily splice the new wires to the old wires at the socket housing, and secure the connection with tape *(inset).* Pull the old wires from the box end of the flexible metal conduit to bring the new wires with them. Leave 8 inches of new wire in the electrical box.
- Untape the old wires and discard them. Twist together the new black wire with the black wire in the box and the new white wire with the white wire in the box. Screw a wire cap onto each connection.

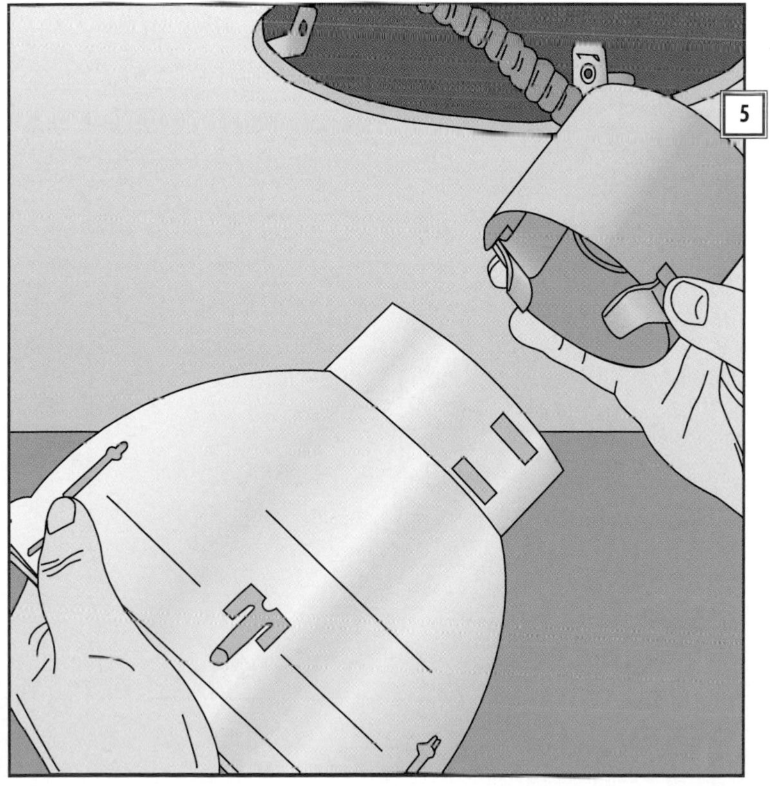



## 5. Reassembling the fixture

- Gently fold the wires into the electrical box, and snap the cover onto the box.
- Squeeze the spring clips, and push the shade into the socket housing *(left).*
- Push the shade into place, and make sure that it is flush with the ceiling.
- Replace the reflector ring, and screw the bulb in. Restore power to the fixture and test it.

# Fluorescent Fixtures

## THREE TYPES OF FIXTURES

The illustrations on this page show typical ceiling-mounted fluorescent fixtures. All work on the same principles and have most parts in common, but there are subtle differences.

In every one, power from a 120-volt household circuit enters the fixture through the top. A ground jumper connects the circuit's ground wire to a ground terminal on the fixture housing. The black and white wires go to the ballast, a transformer-like device that feeds current to the sockets.

In a rapid-start fixture *(top),* the ballast boosts voltage to start the tube, then limits current once the tubes are lit. Older fixtures, like the one in the center at right, employ a component called a starter and two ballast circuits, one to provide the initial surge of voltage and the other for continuous operation. A socket at each end holds a straight tube. Circular tubes plug into a single socket *(bottom).*

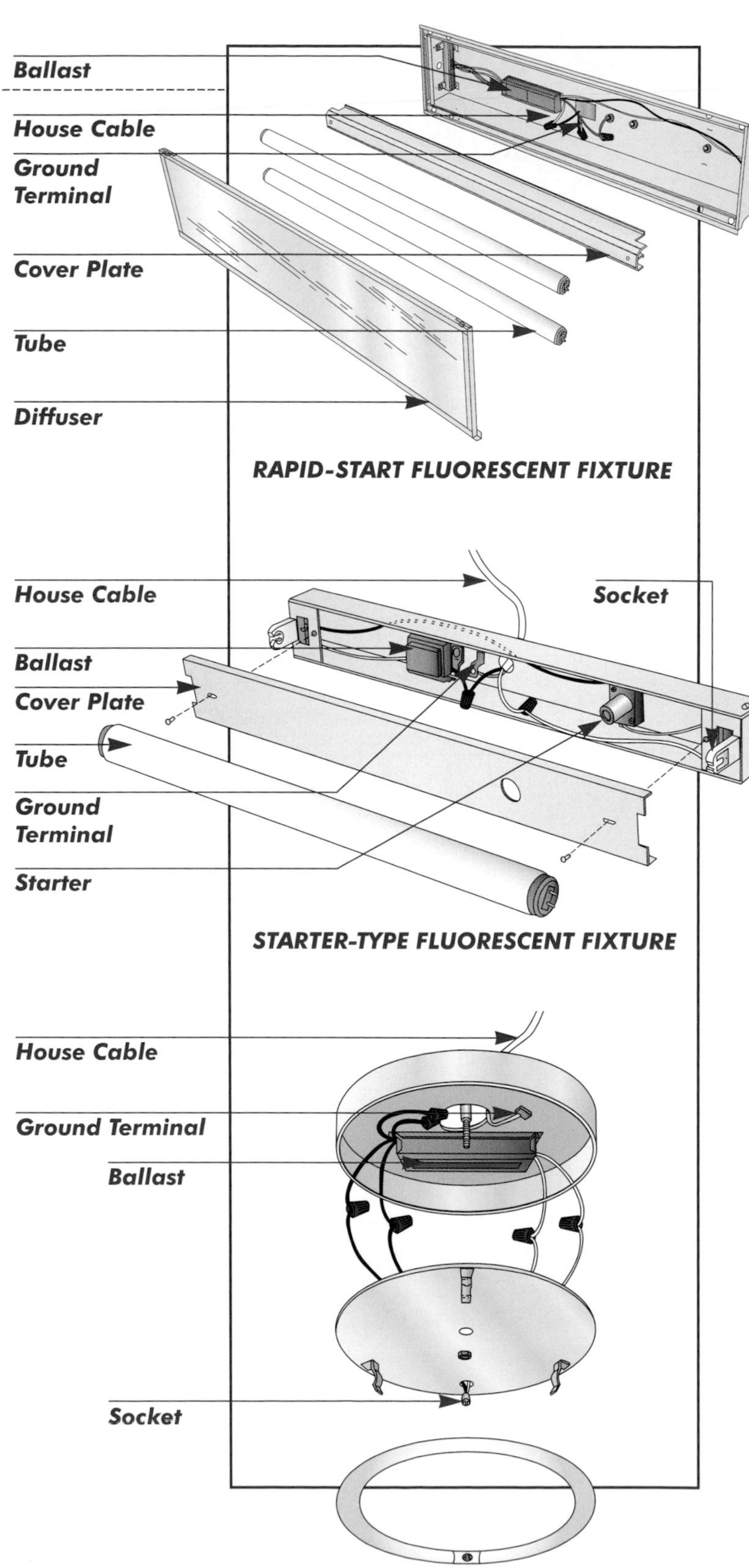


RAPID-START FLUORESCENT FIXTURE



STARTER-TYPE FLUORESCENT FIXTURE



CIRCLINE FLUORESCENT FIXTURE

# Changing Tubes and Starters

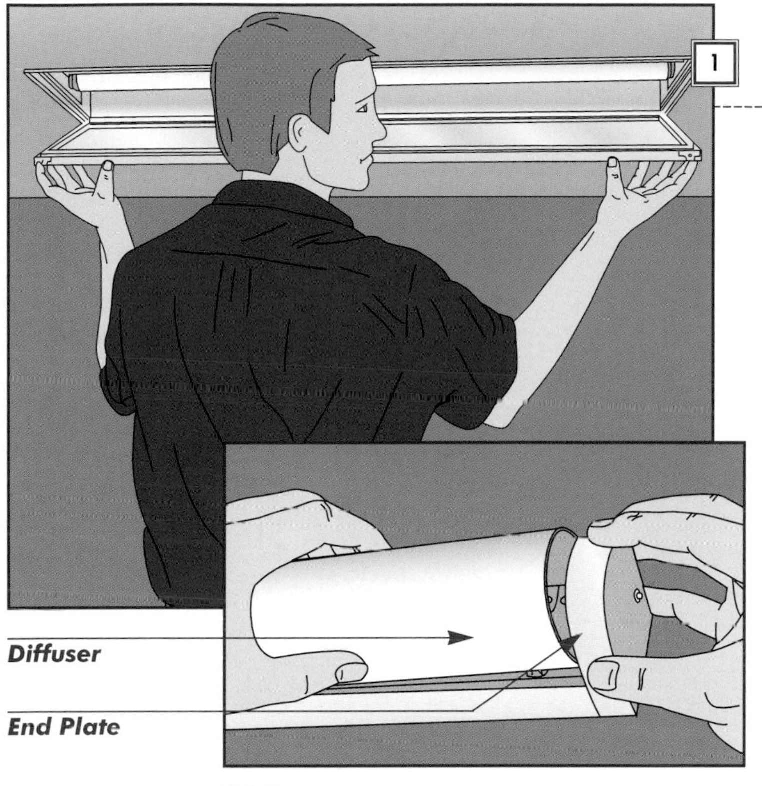


### 1. Gaining access to the tube

• To change a tube, first remove the diffuser to gain access to the tube. There are a number of different designs, but often the diffuser rests in a metal frame hinged with pins on one side and secured by clips on the other. Slide in the clips to release one side, and let the diffuser hang by the pins *(left).*

• On fixtures with a curved, snap-in diffuser, pull one end plate out with one hand while squeezing the diffuser and pulling it free with the other *(inset).*

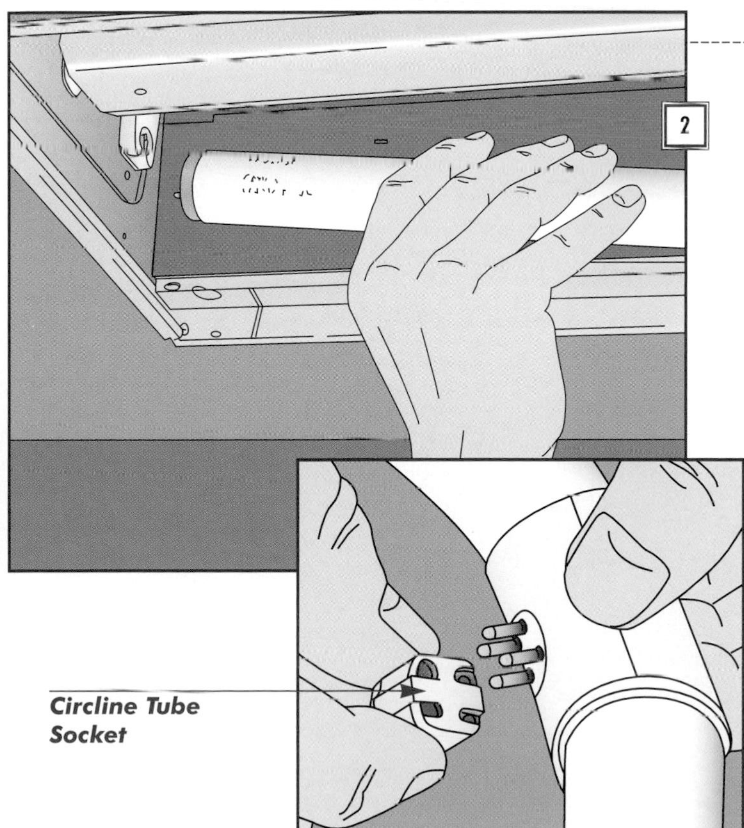


### 2. Removing the tube

• Hold a straight fluorescent tube at both ends; then rotate the tube a quarter-turn in either direction, and lower the pins from the sockets *(left).*

• Detach a circline tube by pulling the socket off the tube pins *(inset)* and releasing the tube from the retaining clips on the fixture.

### 3. Replacing a starter

If the fixture is the rapid-start type, skip this step and go to Step 4.

- Push in the starter and twist it counter-clockwise a quarter-turn; then pull it out *(right).* Take the old starter to a lighting store; the wattage of the new starter must match that of the tube and ballast.

- Insert the new starter so that the pins on one end enter slots in the fixture; then turn the starter clockwise until you hear it click into place.

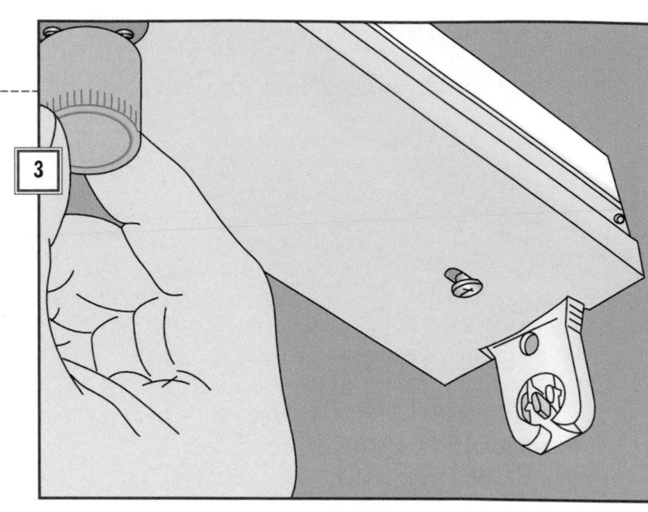


### 4. Installing a new tube

- For a straight tube, insert the pins vertically in the sockets *(right)* and twist the tube a quarter-turn to seat it *(inset).* Replace the socket if the pins do not fit snugly *(page 59).*

- Install a circline tube by lining up the pins with the holes in the socket and pushing the socket onto the pins. Then fit the tube between the fixture's retaining clips.

- Reinstall the diffuser.

## CHOOSING THE RIGHT FLUORESCENT BULB

Information printed at the end of a fluorescent tube *(right)* includes its length, its wattage, and its color temperature—a measure of a light's color expressed in degrees Kelvin (°K): The higher the temperature, the bluer and colder the light.

You must choose a replacement tube of the same length and wattage as the old one (the two are interdependent), but you can select from a wide variety of color temperatures. Tubes rated at 4,000°K and higher produce a harsh light best suited to workshops and other spots were good visibility is crucial.

Tubes rated at 3,000°K or lower produce a warm light comparable to that of incandescent bulbs. If in doubt, choose a medium white bulb (3,000 to 4,000°K); some of these mimic natural light.

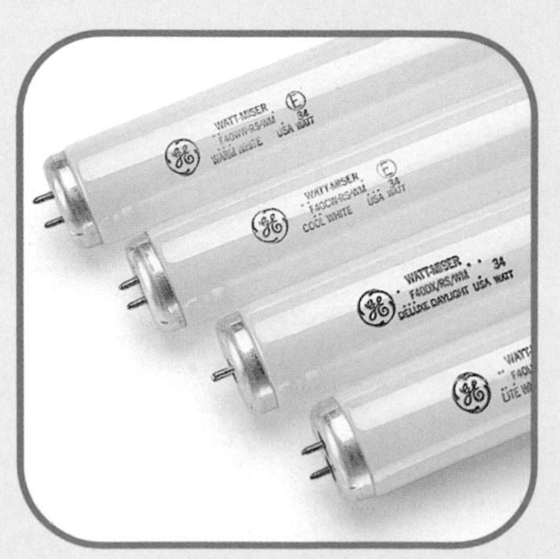

# Ballasts and Sockets

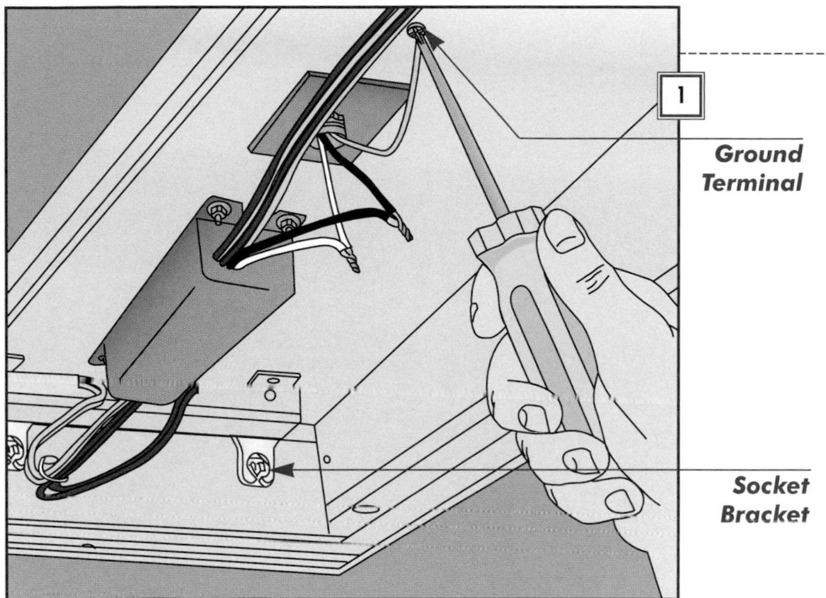


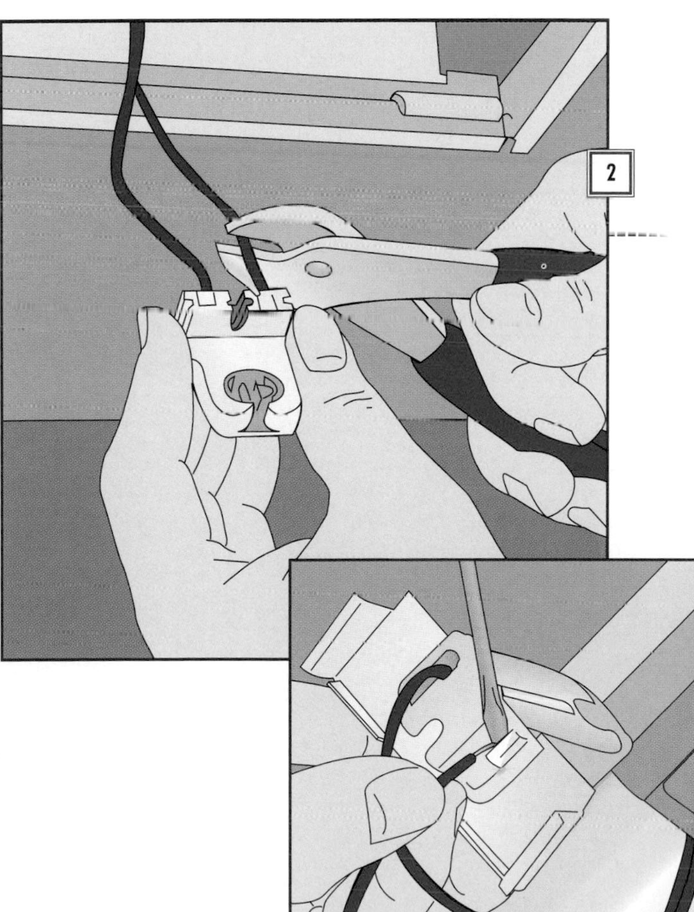


## 1. Testing for Voltage and Checking the Ground Connections

• Flip off the switch and cut power to the circuit at the service panel *(page 10)*. Remove the diffuser and tubes *(page 57)*. Squeeze the cover plate to release it from the frame and expose the wiring.

• Unscrew the wire caps from the black, white, and ground-wire leads, taking care not to touch the bare wire ends.

• Set a multitester to 250 in the AC-voltage range *(page 20)*. Holding the probes by their plastic handles, touch one probe to the ground-terminal screw and the other first to the black wire connection and then to the white wire connection. Next, test between the white and black wire connections. If voltage is present, return to the service panel and cut power to the correct circuit.

• Inspect the ground connection to the fixture. Tighten the screw if it's loose *(left)*.

## 2. Replacing a Socket

• With the power off, unscrew the socket bracket to expose the socket connections.

• If the socket has preattached leads (no push-in terminals or terminal screws), cut the wires close to the socket, leaving long leads *(left)*. Disconnect a socket with push-in terminals by inserting a screwdriver tip into each terminal slot and pulling out the wire *(inset)*. For a socket with screw terminals, loosen the screws and remove the wires.

• Buy a replacement socket with screw terminals. Strip the wire back 1/2 inch on each lead, and connect the wires to the new socket. Either lead can go to either terminal.

• Screw the socket and its bracket to the fixture, then reinstall the tubes and restore power to the circuit. If the light works, reattach the diffuser.

### 3. Replacing a ballast

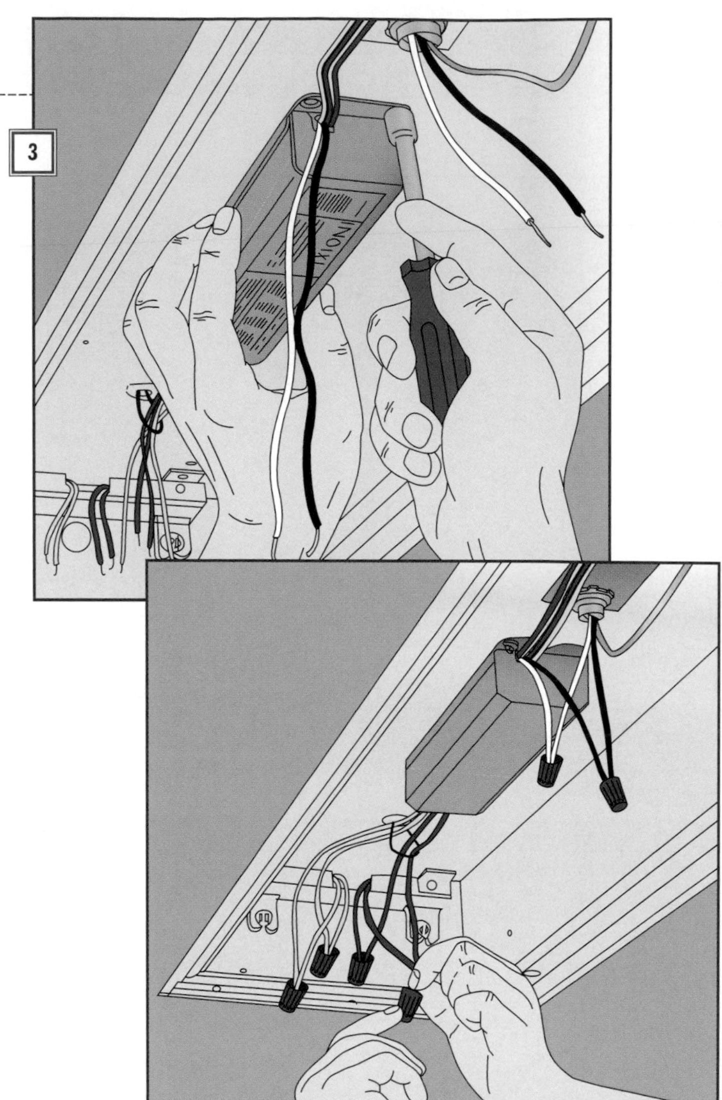


- After turning off the power and checking for voltage *(page 59)*, release the sockets from their brackets and disconnect the ballast from both sockets. Cut the wires near the ballast if the sockets have preattached wires. Disconnect the black and white wire leads.

- Unscrew the ballast *(right)*, and take it to a lighting-supply store for a replacement.

- Fasten the new ballast to the fixture with the black and white leads facing the incoming house wires.

- Connect the ballast to the sockets *(inset)*, and attach the black lead to the black house wire and the white lead to the white house wire, securing the connections with wire caps.

- Set the sockets in place and reattach the cover plate, taking care not to pinch any of the wires or connections.

- Install the tubes, and restore power to the circuit at the service panel. If the light works, reattach the diffuser.

## REPLACE THE BALLAST?

Any of several symptoms may point to a faulty ballast: The fixture won't light; the tube blinks, flickers, or is slow to light; or the fixture hums. The clearest symptom that a ballast is failing is black resin dripping from the ballast casing.

Although a ballast can last up to 12 years, it is generally the most expensive part of a fluorescent fixture. When a ballast does fail, you might want to consider replacing the fixture altogether. Compare the price for a new ballast to the cost for a replacement fixture of the same type. Generally speaking, simple utility fixtures can be had for a price very close to that of the ballast itself. Replacing the ballast may be the better bargain for more expensive, decorative fluorescent fixtures.

# Replacing a Fluorescent Fixture

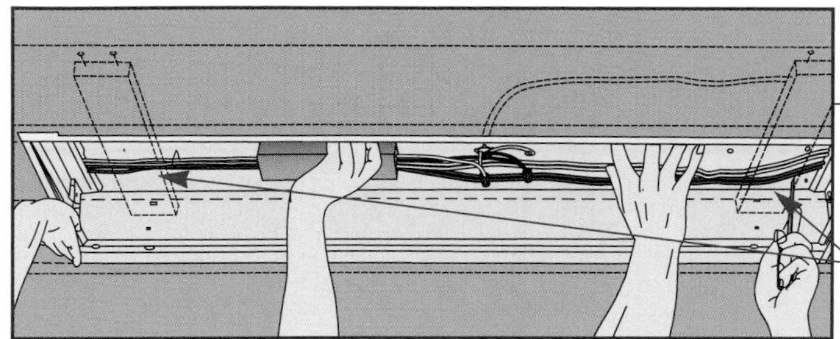


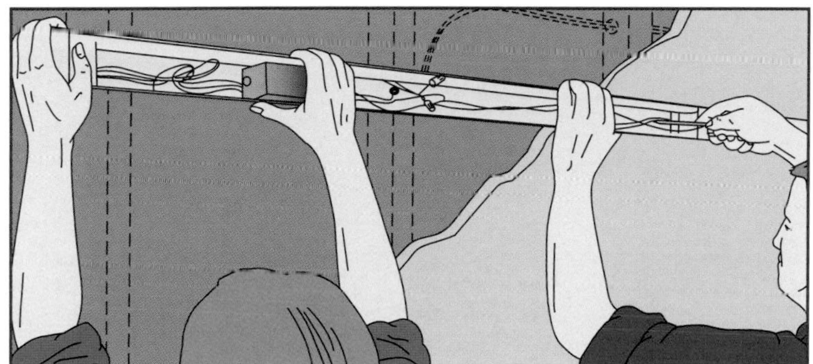

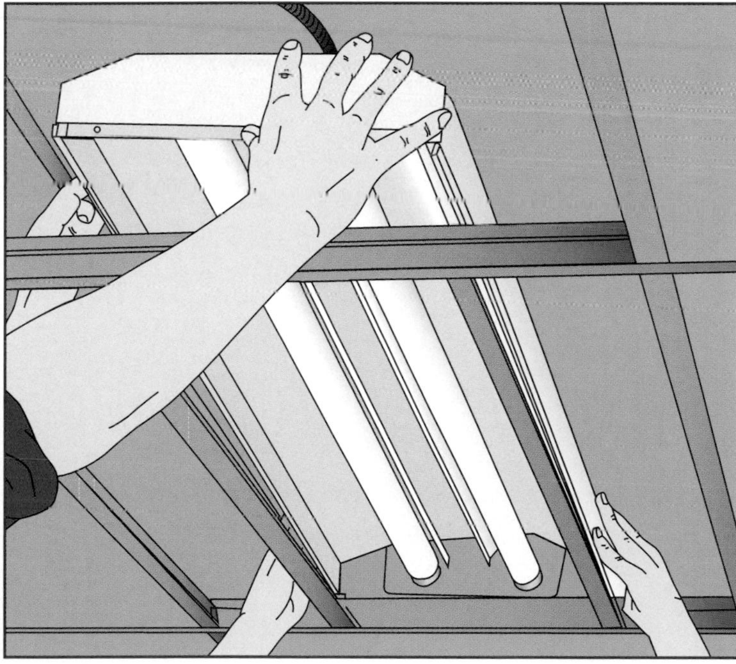

## Dismounting the Old, Installing the New

- Flip on the switch and cut power to the fixture at the service panel *(page 10).*

- Remove the diffuser and tubes. Squeeze the cover plate to release it from the frame, exposing the sockets. Test for voltage to confirm that power is off *(page 59).*

- Twist the wire caps off the connections between the incoming cable and the ballast's black and white leads, then disconnect the cable ground wire from the fixture. Unscrew the locknut on the clamp that secures the power cable to the fixture.

- Have a helper support the fixture while you remove the fasteners. Fixtures recessed in drywall or plaster ceilings are usually screwed to cross braces between joists *(top).* Surface-mount fixtures *(center)* are usually attached directly to the ceiling or wall. Fixtures recessed in a suspended ceiling *(bottom)* are usually supported by the ceiling grid; safety chains, straps, or wire may hold them in place. Remove any ceiling tiles around the fixture for access.

- Have a helper support the new fixture while you clamp the cable to the unit through the knockout in the housing.

- Mount the new fixture to the ceiling.

- Connect the black lead to the black house wire and the white lead to the white house wire. Secure the connections with wire caps.

- Connect the cable's bare copper ground wire to the grounding terminal on the fixture housing *(page 56).*

- Replace the cover plate, install the tubes and the diffuser, and then restore power.

# Ceiling Fans

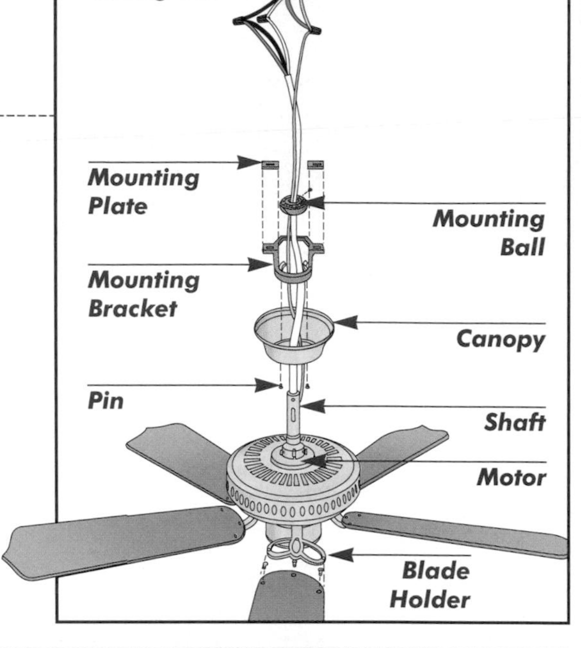


## Ceiling-fan anatomy

A ceiling fan is an electric motor with blades attached *(right).* The assembly is fastened to a ceiling box by means of a mounting plate. A shaft connected to the motor housing extends up and locks into place on a mounting ball; the mounting ball rests in a bracket that supports the fan's weight. A canopy covers the mounting assembly. Many ceiling fans have an integral light fixture with wiring and mounting features similar to those of pendant lights *(page 48).*

## 1. Taking down the fan

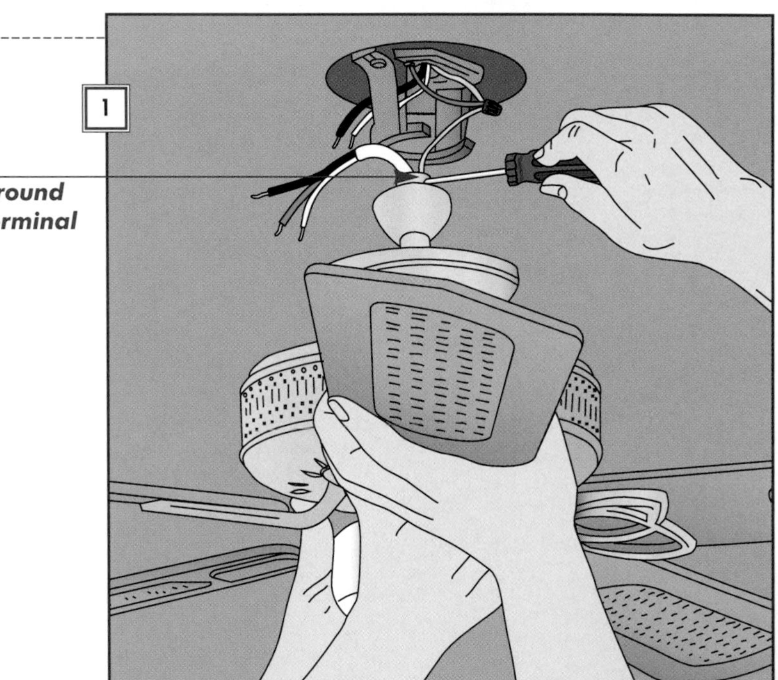


- Flip on the switch and cut power to the circuit at the service panel *(page 10).*
- Loosen the screws holding the canopy, and slide it down the shaft to expose the wire connections and the mounting bracket.
- Pull the black and white wire connections and the blue lighting wire, if any, out of the ceiling box. Unscrew the wire caps from the connections with the black and white cable wires, taking care not to touch any of the exposed wire ends. Leave the ground-wire connections in the box.
- Set a multitester to 250 in the AC-voltage range *(page 20).* Holding the probes by their plastic handles, touch one probe to the mounting plate and the other to the black and white wire connections in turn. The meter should not register voltage. After confirming the power off, disconnect the wires.
- Have a helper raise the fan, then slip the mounting ball out of the bracket. Push the ball downward to expose the ground wire connected to a terminal at the top of the shaft. Loosen the screw *(right),* detach the wire, and lower the fan.

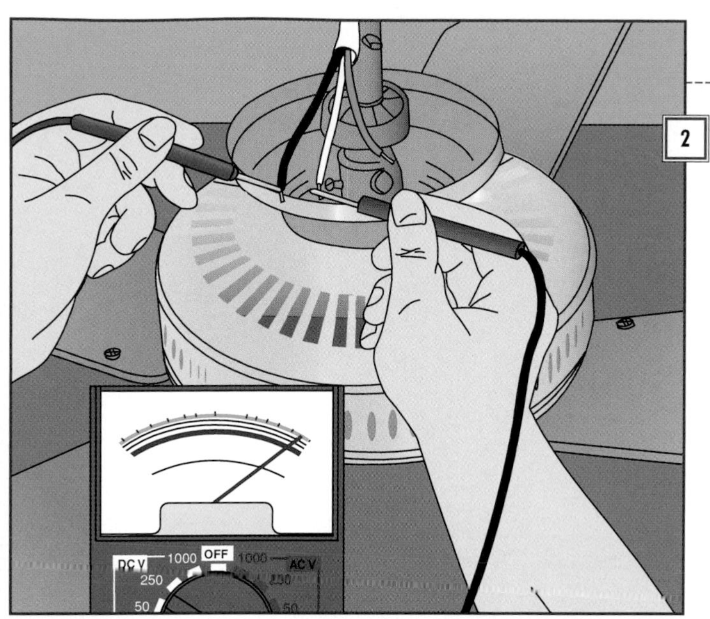


## 2. Testing the motor

• Set a multitester to RX1 and test the motor for continuity *(page 20)*. To do so, set the pull-chain to the ON position and touch one probe to the black lead and the other to the white lead *(left)*. The meter should register between 20 and 30 ohms.

• A higher or lower reading indicates a faulty motor; take it in for service.

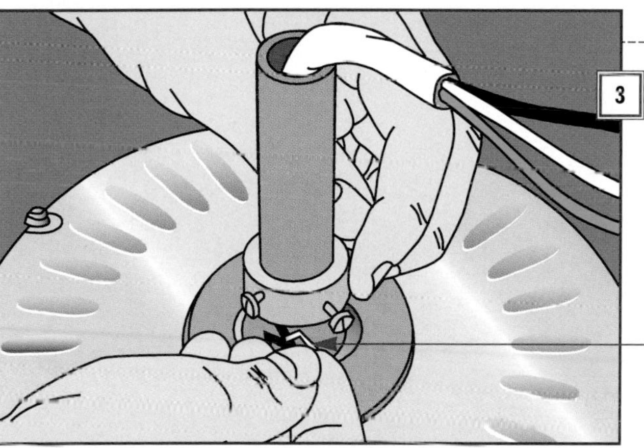


**Cotter Pin**

## 3. Inspecting the pins and blades

• Pull out the cotter pin that secures the motor housing to the fan shaft *(left)* and the pin that holds the mounting ball in place; inspect each pin. Replace either pin if it appears worn.

• Check the blades and blade holders. If any blades are warped or broken, replace the entire set. Tighten any loose blade screws.

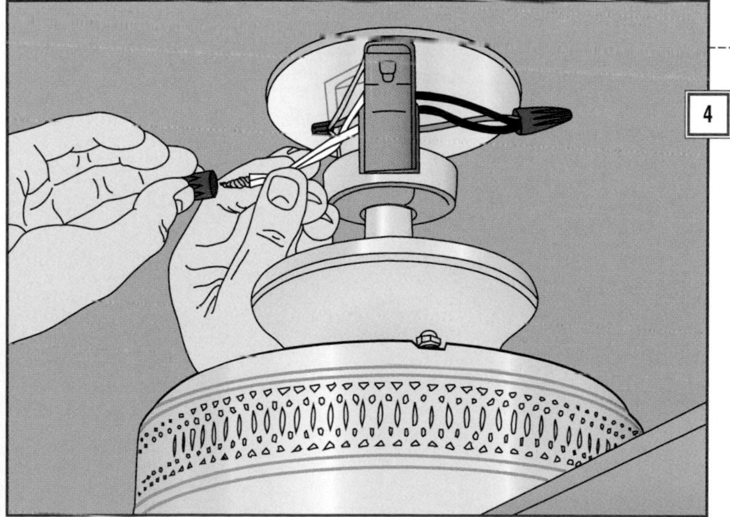


## 4. Reinstalling the fan

• Reconnect the ground wire to the terminal at the top of the shaft. Slide the mounting ball up the shaft, and slip the ball into its slot in the mounting bracket.

• Twist the black fan and the black house wires together along with the blue wire (if the fan has a light), and connect the white fan wire with the white house wire. Screw wire caps on the connections *(left)*, and fold the wires back into the ceiling box.

• Slide the canopy up the shaft, and tighten the screws that hold it in place. If the canopy tends to mar the ceiling, lower it 1/4 inch so that it no longer rubs.

• Restore power to the circuit.

# FIX IT: Wall Switches

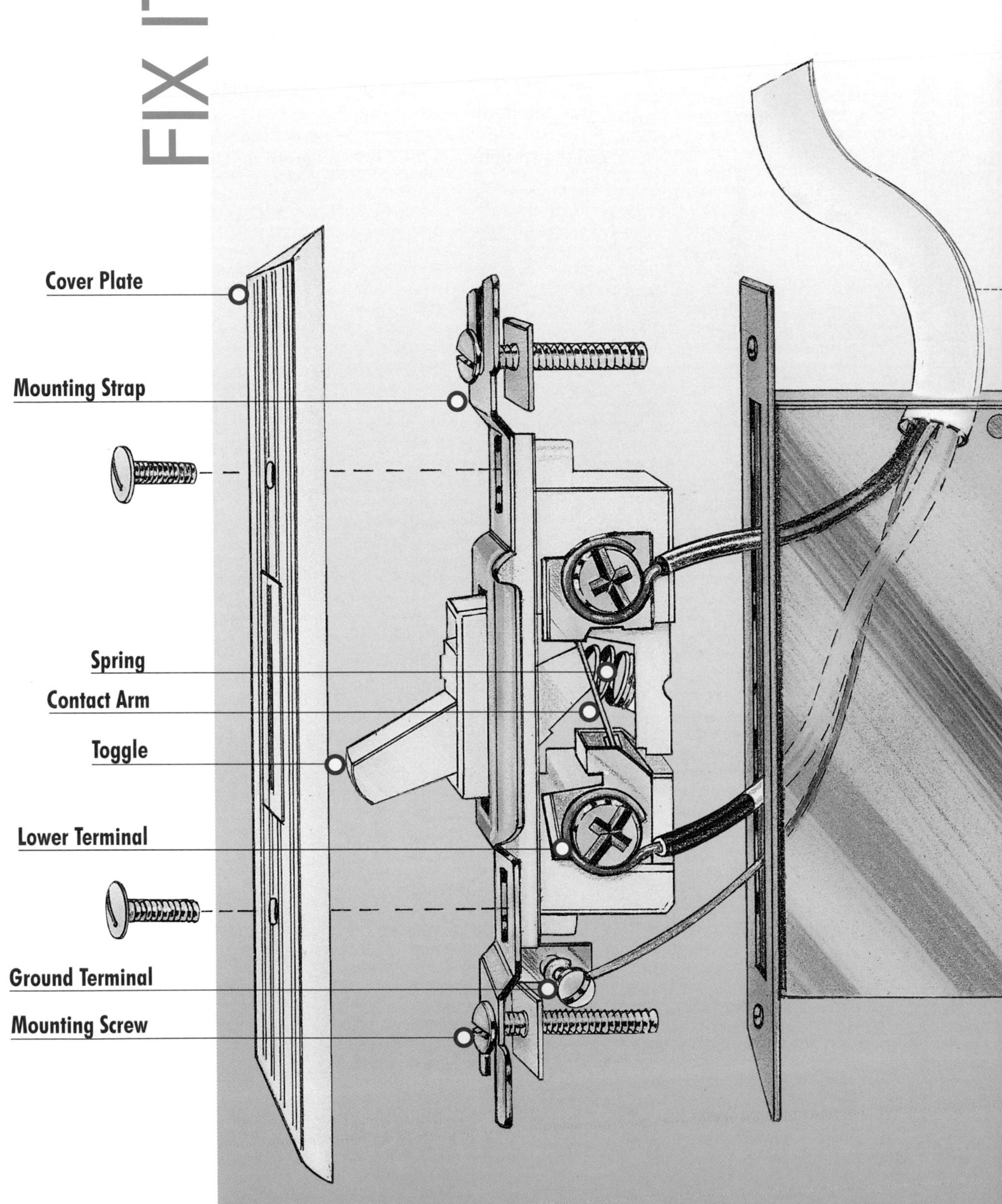

# Chapter 4

## How They Work

**A single-pole switch *(left)* has only one pair of terminals and can complete the circuit only in one position. Inside a typical switch is a metal bar called a contact arm. It's hinged at the bottom, where it's always in contact with the lower terminal. When the switch is in the OFF position, as it is here, a spring holds the top of the bar away from the upper terminal. Flipping the toggle ON snaps the bar into contact with the upper terminal, completing the circuit.**

**There are several switch variations: The terminal configurations of 3- and 4-way switches *(pages 71 and 74)* enable them to control lights from more than one location. Dimmer switches *(page 75)* offer infinite adjustability to control the intensity of light. A special device combines a switch with an outlet *(page 79)*, which can work together or independently. A pilot-light switch *(page 82)* signals when a light in an unseen area is on. Timer switches *(page 83)* turn lights on and off automatically, according to preset schedules.**

## Contents

# Troubleshooting

| Problem | Solution |
| --- | --- |
| • Switch-controlled light won't turn on | Replace bulb • Replace fuse or reset breaker **10** • Check switch-wiring connections **68–69** • Test and replace switch **68–70** • Check the light fixture **Chapter 3** • |
| • Fuse blows or breaker continually trips when you turn on a switch | Adjust load on circuit **13** • Check switch-wiring connections **68–69** • Check the light fixture **Chapter 3** • |
| • Light flickers, or the switch sparks or makes a crackling noise | Check switch-wiring connections **68–69** • Test and replace switch **68–70** • |
| • Toggle does not stay ON | Replace switch **70** • |
| • Both 3-way switches must be flipped to turn on fixture | Check switch-wiring connections **68–69** • Test switches and replace if faulty **71–72** • |
| • Fixture served by 4-way switch does not light | Check switch-wiring connections **69** • Test switches and replace if faulty **74** • |
| • Dimmer turns light on and off but will not dim | Replace dimmer **75–76** • |
| • Pilot-light switch does not work | Test and replace switch **82** • |
| • Timer switch does not turn fixture on at the preset time | Test and replace switch **83** • |

# Before You Start

## When a light controlled by a switch doesn't work, begin your investigation by replacing the bulb, resetting the circuit breaker, or changing the fuse.

### CHECK THE SWITCH LAST

Sometimes problems crop up in circuit wiring; flip on the other switches and test a lamp in the outlets on the circuit in question. Or the difficulty may lie in the wiring connections to the switch; tighten them before replacing the switch.

After eliminating these potential problems, *then* look to the switch. Most wall switches last 20 years or longer; and when they finally wear out, it's the mechanism that goes, rather than their ability to conduct electricity. In some instances, you may choose to replace a switch in good working order with one of a different kind. Dimmer switches, for example, are handy in many parts of the house. Other switches can turn lights on and off automatically or even tell you when you've left the cellar light on.

### Before You Start Tips:

- Many switch problems can be solved by simply tightening the connections.
- If you must replace a switch, buy a new one that matches the specifications of the original (*page 71*).
- Replacing a switch is a simple matter of moving the wires from the old switch to the corresponding terminals of the new switch.

Multitester
Utility knife
Wire stripper or multipurpose tool
Diagonal-cutting pliers
Long-nose pliers
Lineman's pliers
Screwdrivers

Wire caps
Electrical tape
Fine sandpaper

Before working on a switch, turn off power to the circuit by removing the fuse from the service panel or tripping the circuit breaker. Never restore power until you have returned the switch to its box.

# Testing and Replacing a Single-Pole Switch

## 1. Removing the Cover Plate

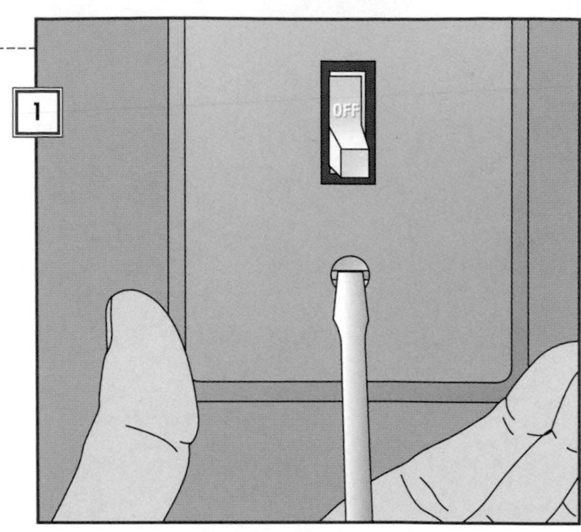


- Cut power to the circuit at the service panel by switching off the circuit breaker or removing the fuse *(page 10).*

- Take out the two screws securing the cover plate *(right),* and lift it off. If the plate is stuck to the paint, cut neatly around its edge with a utility knife.

- Tape the screws to the cover plate so you won't lose them.

## 2. Checking for Voltage

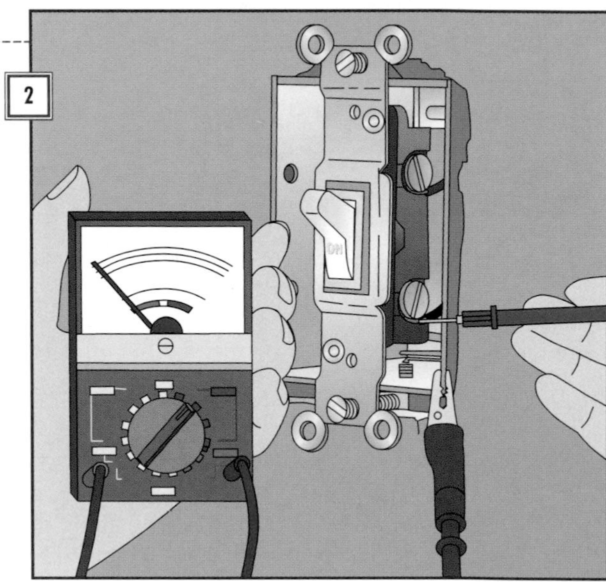


Before proceeding, confirm that power to the circuit is off, as follows:

- Set a multitester to 250 in the AC-voltage range *(page 20).* If the wall box is metal, touch one tester probe to the box; for plastic boxes, touch the probe to a bare grounding wire. In both cases touch the other probe to each wire connected to the switch *(right).* The multitester should register 0 voltage in each case.

- If any reading is greater than 0, return to the service panel and turn off power to the circuit.

*Because of a possibility of current at the switch, hold multitester probes by their plastic grips.*

## 3. Freeing the Switch

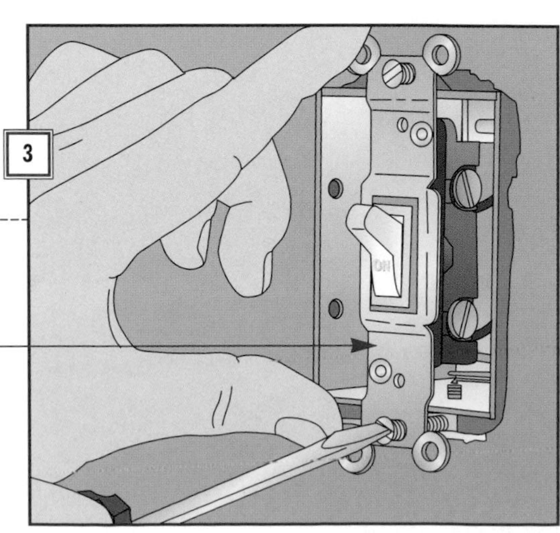


***Mounting Strap***

- Remove the mounting screws at the top and bottom of the mounting strap *(right).*

- Grasp the mounting strap, and pull the switch from the box.

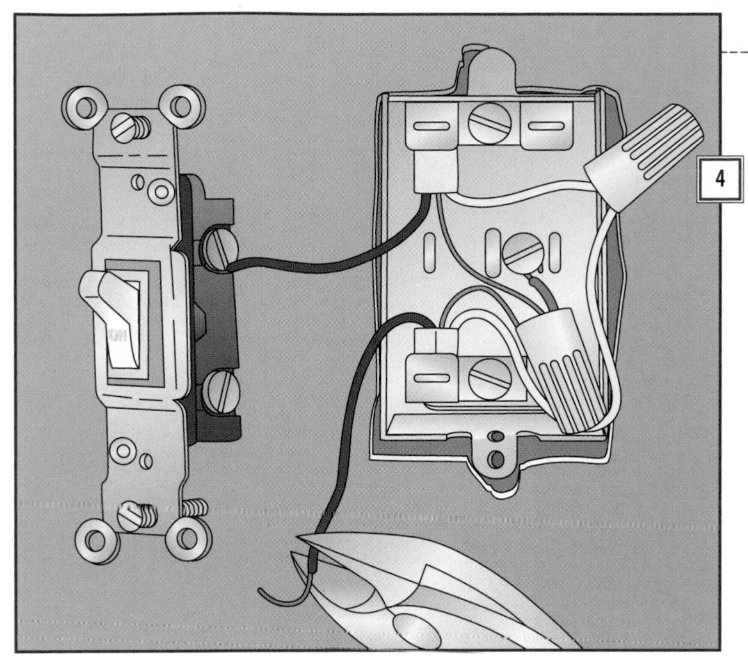


### 4. Checking terminal connections

- Use a screwdriver to tighten any loose connections.
- To clean dirty connections, unhook the wires and brighten the ends and terminals with fine sandpaper *(page 89).*
- If the wire ends are burned or damaged, clip them off *(left),* restrip the ends *(page 110),* and form new hooks *(below).*
- Reconnect the wires, return the switch to the box *(page 70),* turn on the power, and flip on the switch.
- If it works, reattach the cover plate. If not, shut off power to the circuit again *(Step 2),* then proceed to Step 5.

## MAKING PERFECT CONNECTIONS

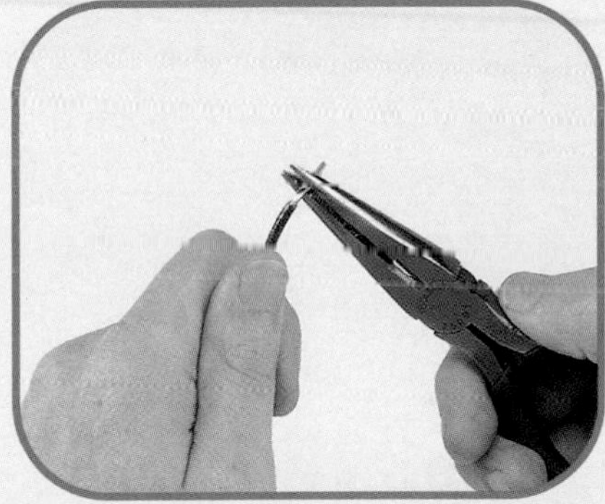

Poor connections can cause lights to flicker and dim, and electric motors to run hotter and die sooner. They have even been known to overheat wiring and start fires. A switch may have wire leads that are connected to the house wiring with wire caps *(page 73),* or wires may connect to the switch at screw terminals.

To create a hook for connecting a wire to a screw terminal, first strip about ½ inch of insulation from the wire end. Using long-nose pliers , make a right-angle bend *(above, left).* Move the pliers' jaw tips about halfway along the stripped wire end, and bend it about 45 degrees in the opposite direction *(above, center).* With the tips of the pliers, bend the wire to a C-shape, leaving the loop open to fit around the screw *(above, right).*

## 5. Disconnecting the switch

• Remove the screws securing the switch in the box, and pull out the switch.

• Loosen the terminal screws and disconnect the wires to free the switch *(right).*

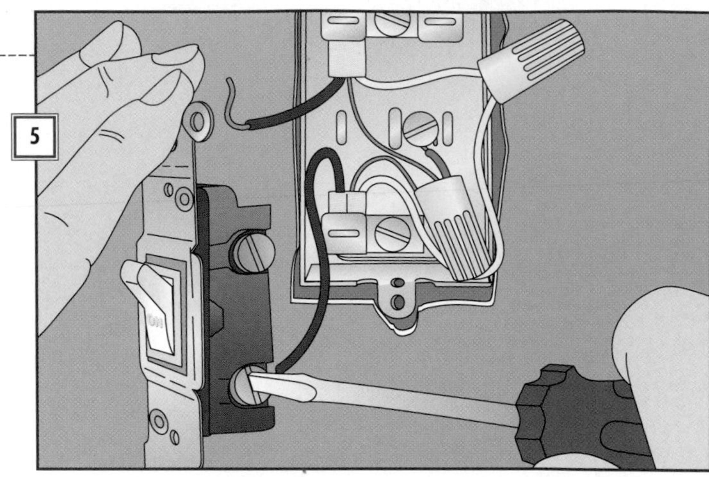


## 6. Testing the switch for continuity

• Set a multitester to RX1, turn on the switch, and clip one probe to each switch terminal *(right).* The meter should indicate continuity *(page 20).*

• Turn off the switch and repeat the test. The meter should indicate an open circuit.

• If your results differ, replace the switch.

## 7. Connecting the new switch

• Hold the switch so the toggle points up when ON.

• Hook the wires around the terminal screws of the new switch just as they had been connected to the old one. Tighten the screws securely.

• Put the switch in the box, carefully folding the wires to make them fit *(right).*

• Screw the mounting strap to the box, making sure the switch is straight. Reattach the cover plate and turn on the power.

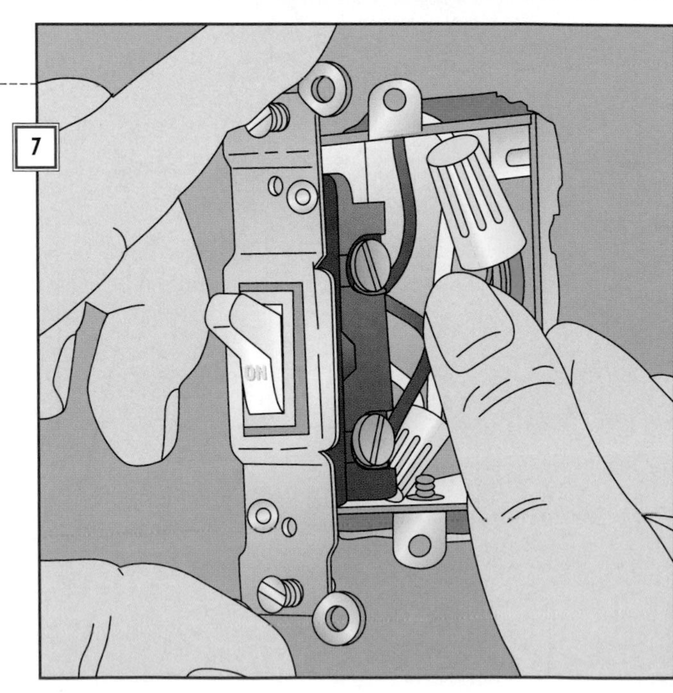

# Troubleshooting a 3-Way Switch

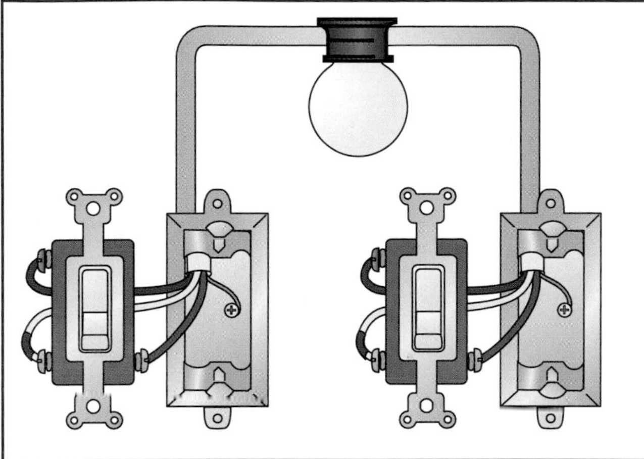

### A 3-WAY CIRCUIT

Named for the number of terminals per switch, 3-way switches control a light from two locations. The illustration at left shows a typical wiring scheme, in which black wires connect to switches' common terminals (black or copper), white wires are recoded as black with tape, and red wires connect to the switches' brass terminals.

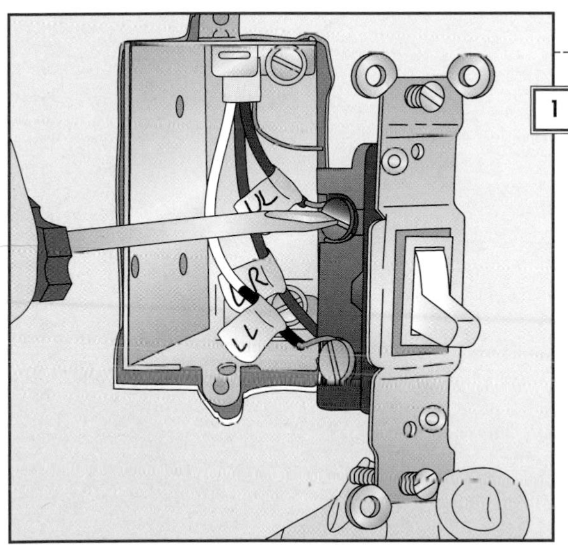

### 1. REMOVING THE OLD SWITCH

- Turn off power to the circuit at the service panel *(page 10)*.
- Remove the cover plate and, with a multitester, check for voltage at the switch terminals *(page 68)*.
- Remove the mounting screws, and pull the switch from the box.
- Tag each wire with a label indicating the switch terminal to which it is connected.
- Loosen the terminal screws and pull off the wires *(left)*.

## READING A SWITCH

Wall switches have important information stamped on both sides. The mounting strap is stamped with maximum amperage and voltage ratings *(right, top):* A 15A-120V switch can handle 15 amps and 120 volts. The testing agency should also be listed. Look for UND LAB INC LIST (Underwriters Laboratories) or for CSA (Canadian Standards Association). On the back *(right, bottom)*, look for wire sizes: Lighting circuits use No. 12 or 14 wire.

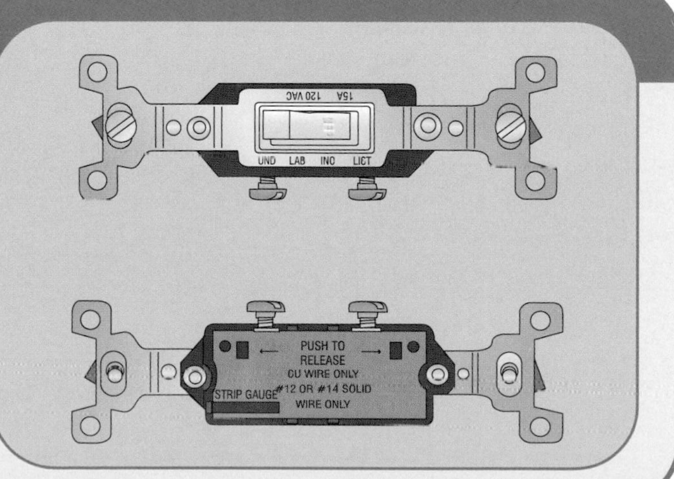

## 2. Testing a 3-way switch

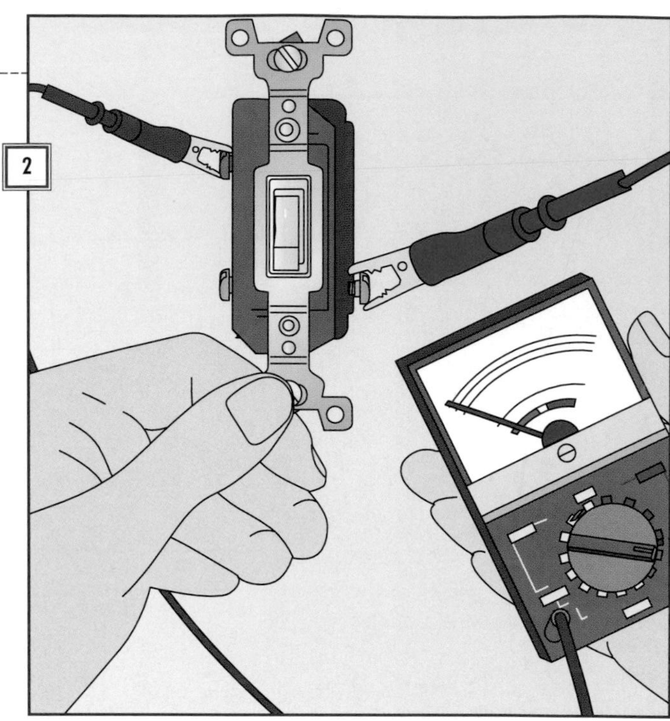


- Set a multitester to RX1. Clip one probe to the black or copper terminal, and the other probe to either of the others *(right).*

- Toggle the switch. If the switch is good, the meter will indicate continuity in one position and not the other *(page 20).*

- Set the toggle to the position in which the meter indicated continuity in the previous test, then clip the black probe to the terminal you haven't yet tested. The meter should indicate no continuity.

- Without moving the probes, toggle the switch. The meter should indicate continuity.

- If the first switch passes, check the other one in the circuit.

- If either switch fails, replace it *(Step 3).*

## 3. Installing the new switch

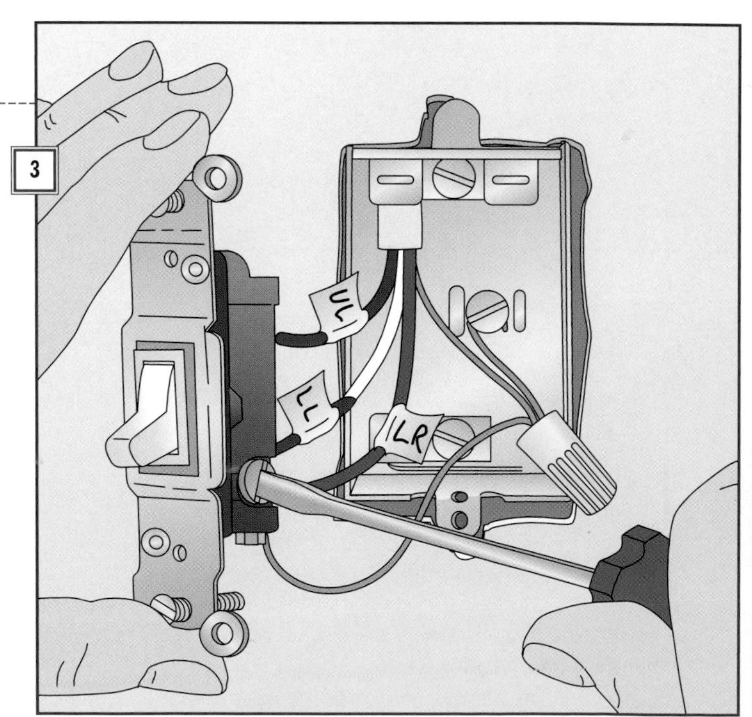


- Connect the wires to the terminals of the new switch as they were connected to the old one, using the tags you attached to them as a guide *(right).* As a rule, the incoming hot wire (black) goes to the black or copper terminal on the switch. Other wires—white, white recoded as black with electrical tape, and red—are generally connected to the brass or silver terminals. If present, the bare copper or green wire should be connected to the green ground terminal. Positions of terminals on the switch vary with the manufacturer.

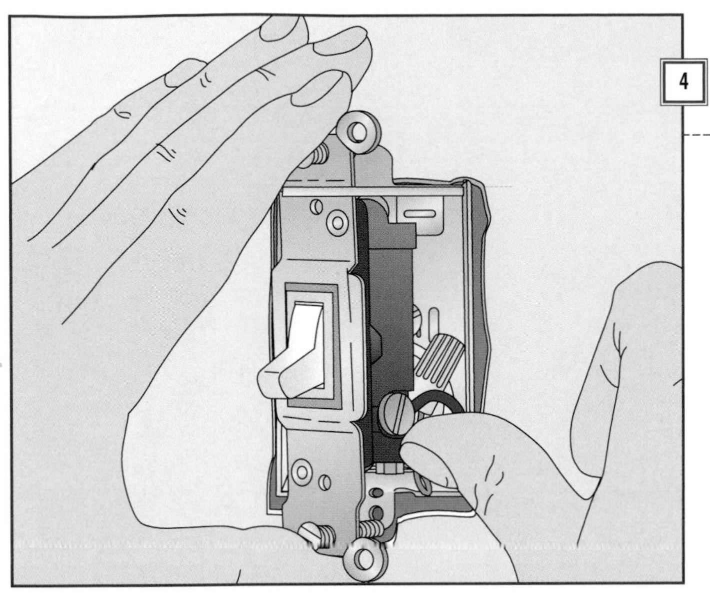


### 4. Remounting the switch

- Fold the wires into the switch box while pushing the switch into place *(left).*
- Tighten the mounting screws to fasten the switch to the box.
- Remount the switch plate, and turn on power to the circuit at the service panel.

## Ron's TRADE SECRETS

**Making tight splices**

When I splice wires, I like to make sure they can't come apart. So I usually strip off about 1 full inch of insulation from the ends of the wires to be joined—more than actually seems necessary. I hold the wires side by side with one hand, grip the bare ends with a pair of pliers, and twist them together clockwise until the turns are tight and uniform *(right, top).* Then I test the connection with a slight tug.

Next I clip about 1/2 to 3/4 inch off the splice with diagonal-cutting pliers, snipping an angle to create a distinct point at the spliced wire ends *(right, center).* The point ensures that the wire cap threads tightly over the wires as I twist it on *(right, bottom).* I tighten the cap down clockwise until no copper shows.

Finally, I wrap electrical tape around the base of the cap, then once or twice around the wires to make sure that everything stays put.

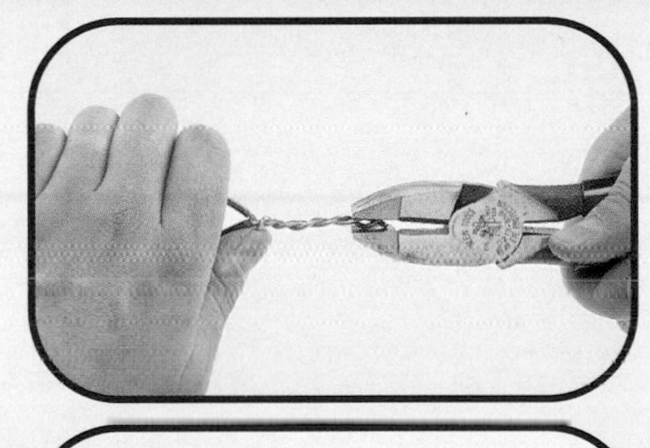

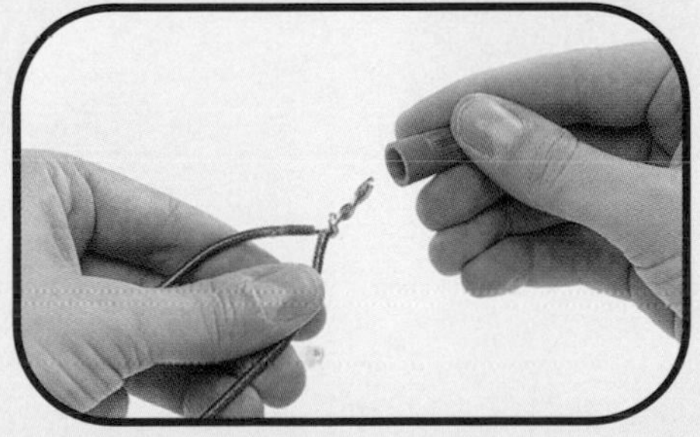

# Locating a Faulty 4-Way Switch

## A 4-WAY CIRCUIT

Installed between a pair of 3-way switches, a 4-way switch *(right)* lets you control a fixture from three or more locations. Four-way switches have two copper and two brass screw terminals. There are two three-wire cables entering the box. The red wires are connected to one side of the switch, the white wires to the other. The black wires are joined with a wire cap.

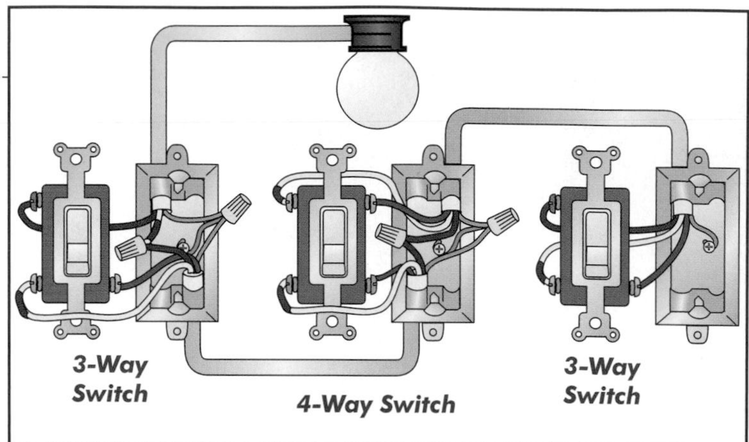


## 1. TAGGING THE WIRES

- Turn off power to the circuit at the service panel *(page 10);* then remove the cover plates from all three switches and test for voltage *(page 68).*

- After confirming that the power is off, free all switches from their boxes *(page 68).* Tag the two wires connected to the upper terminals of the 4-way switch before you disconnect all wires from the terminals.

- Test the 3-way switches on the circuit for continuity *(page 72).* If neither is faulty, proceed to Step 2.

1

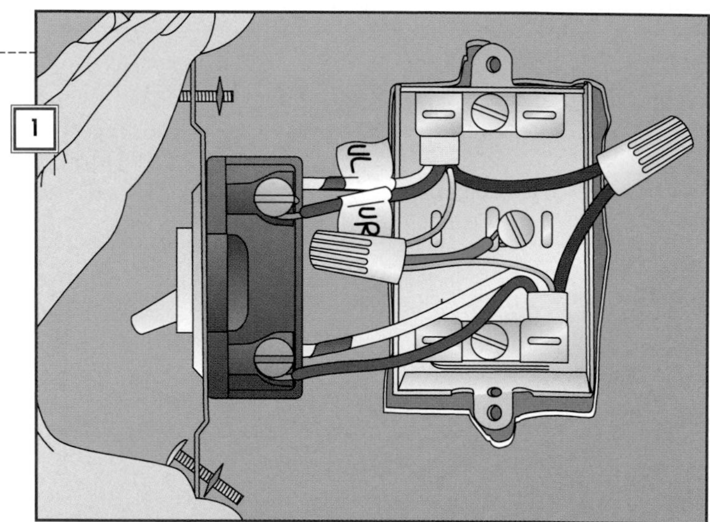


## 2. TESTING THE 4-WAY SWITCH

- Clip multitester probes to the terminals on the right side of the switch *(right).* The meter should show continuity with the switch in one position but not the other.

- Keeping one probe on the upper right terminal, place the other on the lower left. The meter should indicate continuity with the toggle in the opposite position.

- Place the probes on the left-side terminals. You should see a continuity reading on the meter when the toggle is opposite where it was in the first test.

- If the switch fails any test, replace it.

2

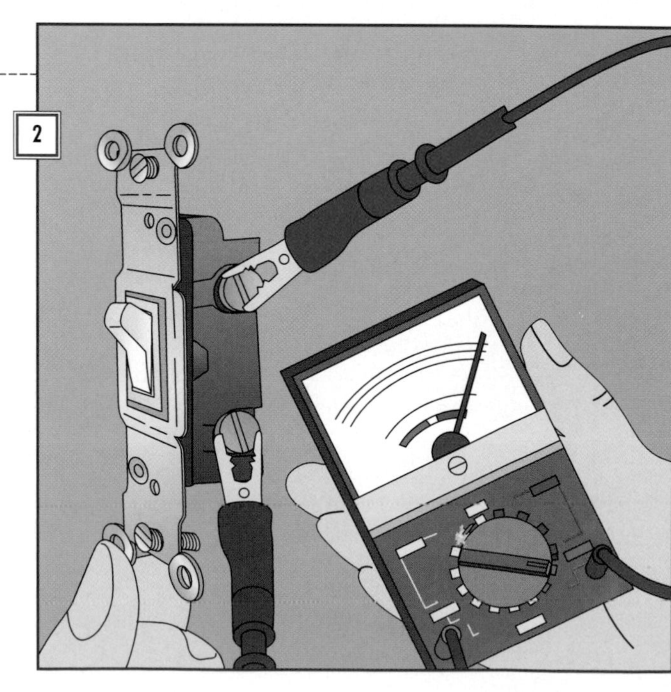

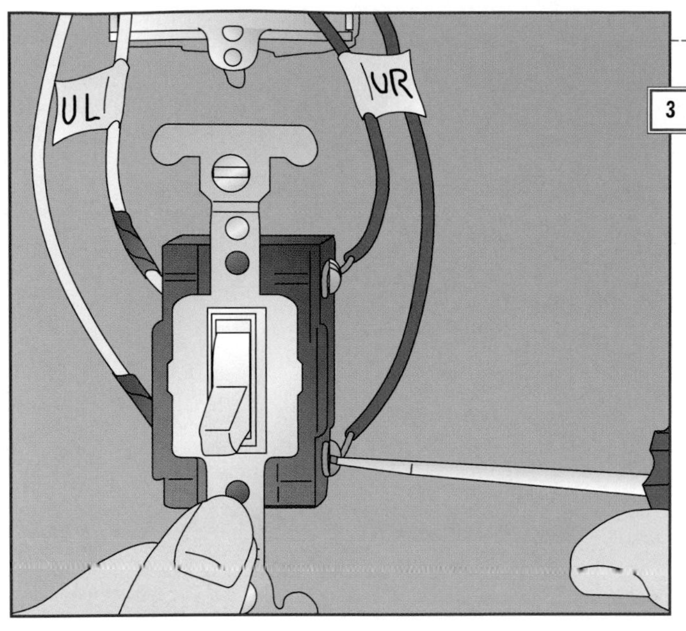


### 3. Reconnecting the switch

• Using the tags you placed on the wires as a guide, reconnect the wires to the switch terminals exactly as before *(left).*

• Press the switch into the box and fasten it with the mounting screws. Attach the cover plate, and restore power to the circuit at the service panel.

## Replacing a Dimmer Switch

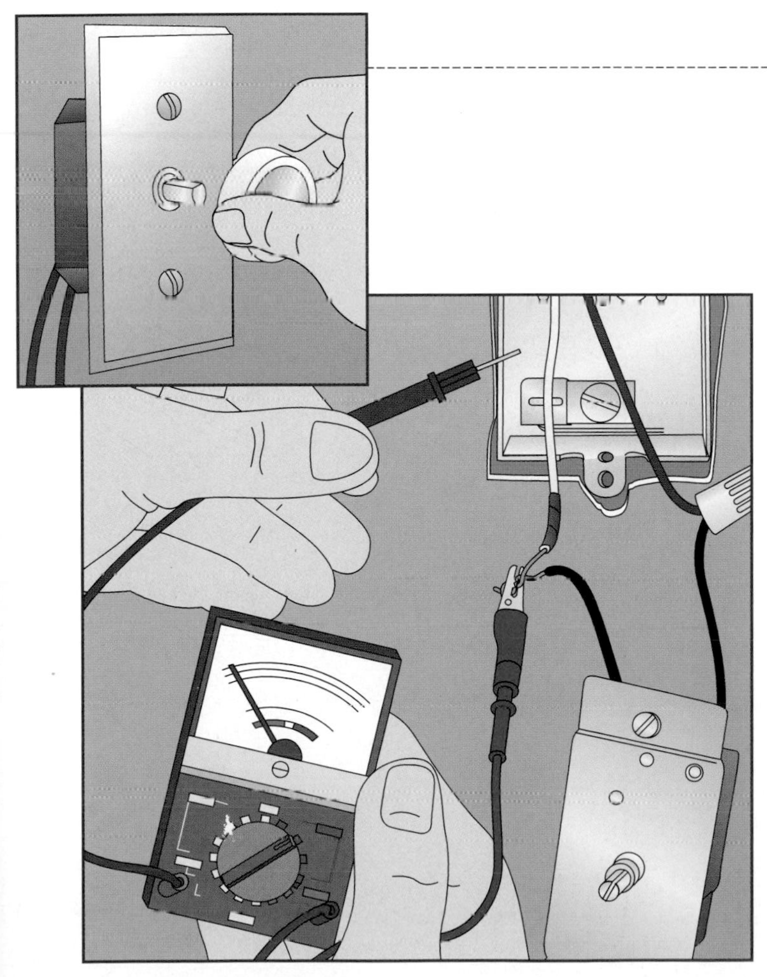

### Checking for voltage

• Turn off power to the circuit at the service panel *(page 10).*

• Pull off the control knob *(upper left)*, and unscrew the cover plate. Free the switch housing *(page 68).* Remove the wire caps joining the switch leads to the house wires.

• Set a multitester to 250 in the AC-voltage range. Touch one probe to the box if it's metal (or ground wire if it's plastic) and the other to each exposed splice in turn. Then test the wire connections against each other. If the meter shows voltage present, return to the service panel and cut power to the circuit before continuing.

• Check for loose connections, corrosion, and burned wires at the splices. Clean the wire ends or clip the wires back *(page 89)*, as necessary. Then reinstall the dimmer and turn on the power.

• If the switch doesn't work after these efforts, replace it *(next page).*

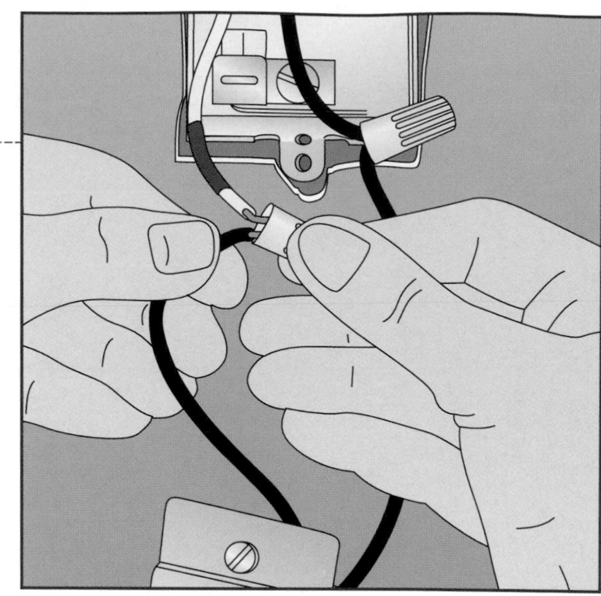

## CONNECTING A SINGLE-POLE DIMMER

Single-pole dimmer switches have two leads. Get a replacement rated at the same wattage as the old switch.

• To install a single-pole dimmer, secure either lead to either wire in the electrical box with wire caps *(right).*

• Place the switch in the box, attach the cover plate and control knob, and turn on the power at the service panel.

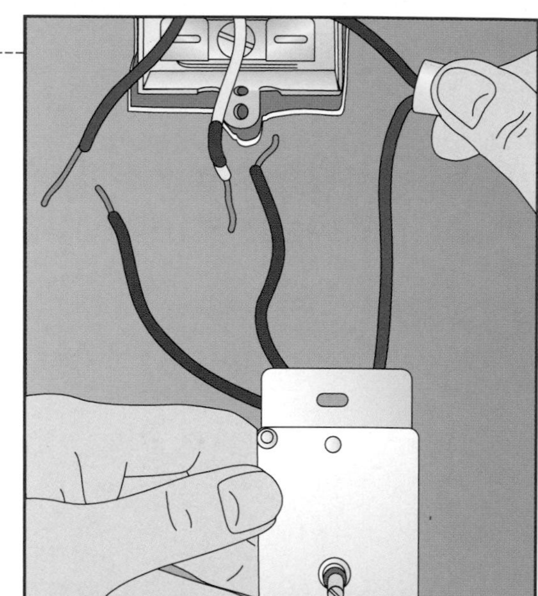

## CONNECTING A 3-WAY DIMMER

Three-way dimmer switches have three leads. Buy a replacement rated at the same wattage as the old switch.

• Using a wire cap, attach the dimmer's red lead to the circuit's black wire *(right).*

• Attach each remaining lead to one of the remaining circuit wires with wire caps.

• Place the switch in the box, attach the cover plate and control knob, and turn on the power at the service panel.

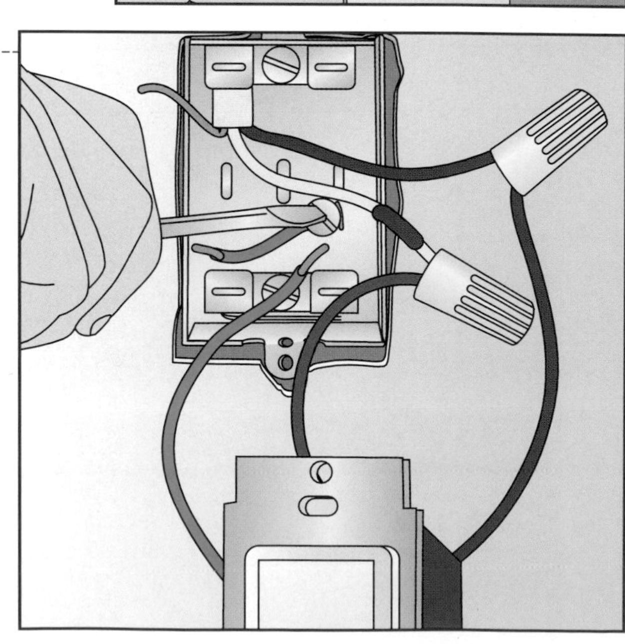

## CONNECTING A TOUCH-SENSITIVE DIMMER SWITCH

Some dimmers have a touch-sensitive panel instead of a knob. These generally have three leads: a black, a red, and a green.

• Attach the black lead to the circuit's black wire, and the red lead to the white wire.

• Connect a green jumper to the ground terminal in the box *(right).* Then, using a wire cap, join the jumper to the dimmer's green wire and to the ground wire in the cable.

• Place the switch in the box. Attach the cover plate, and turn on the power at the service panel.

# Servicing Ganged Switches

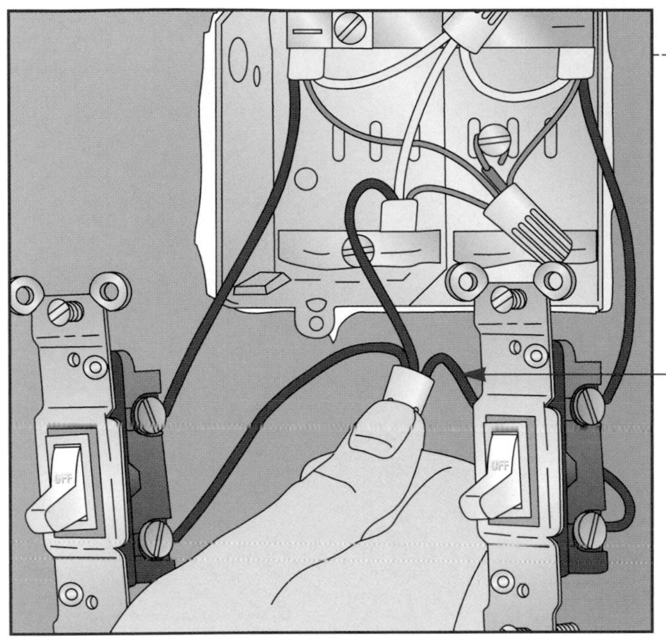


### Multiple switches on the same circuit

When two switches occupy one box, both are on the same circuit if one black wire is connected to both switches by jumpers *(left)*.

• Shut off power to the circuit at the service panel *(page 10)*.

• Remove the cover plate, and test for voltage *(page 68)*.

• When you are sure that power to the circuit is off, unscrew the wire caps as necessary and inspect all connections *(page 69)*. Clean terminals and the wires, clip off any burned ends, and restrip the wires; then reconnect them as you found them.

• Return the switches to the box, turn on the power, and flip the switch.

• If the fixture does not light, turn off the power, disconnect the switch, and test it for continuity *(pages 20 and 70)*.

• If the switch fails the continuity test, replace it with a new one.

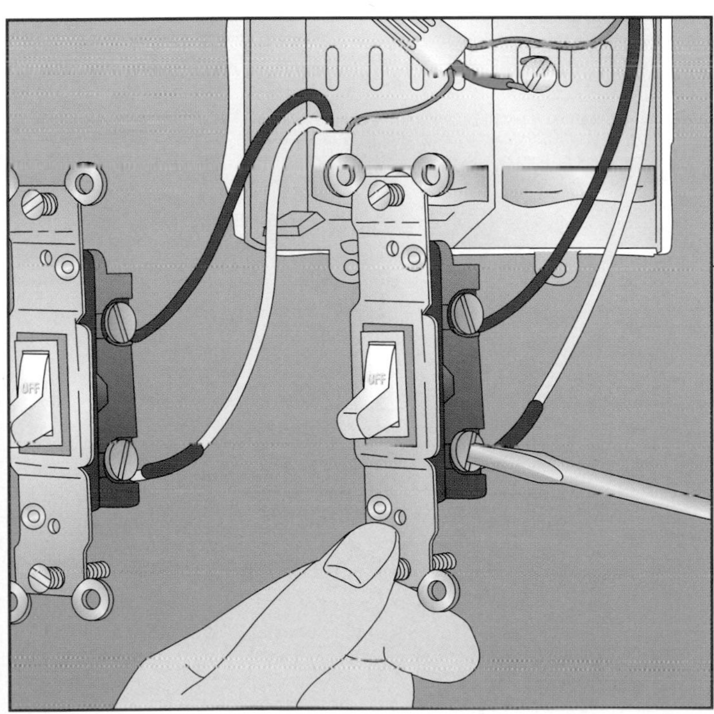

### Multiple switches on separate circuits

Two switches installed in a single box are on different circuits if none of the wires are connected to both switches by jumpers.

• Shut off power to both circuits at the service panel *(page 10)*. Perform voltage tests *(page 68)* to confirm that power to both circuits is off.

• Disconnect the inoperative switch *(left)*. Check the connections, and clean or repair as necessary *(page 69)*; and with a multi-tester, test the switch for continuity *(page 70)*. Replace the switch if it fails the test.

# Testing and Replacing a Combination Switch

## 1. Tagging the Wire to the Connecting Tab

A combination switch has two toggles *(right).* If the switch has one black wire connected to a terminal linked to another by a connecting tab *(inset),* power comes to the switch from a single source. If the connecting tab has been removed and two black wires are attached to the terminals, the switches are on separate circuits.

- Turn off power at the service panel; then remove the cover plate and test both switches for voltage *(page 68).*

- Before detaching any wires *(page 70),* tag the wire attached on the side with the connecting tab.

- Check and repair the connections as needed *(page 69);* return the unit to the box; restore power; and turn on both switches.

- If either doesn't work, turn off the power and remove the switch from the box.

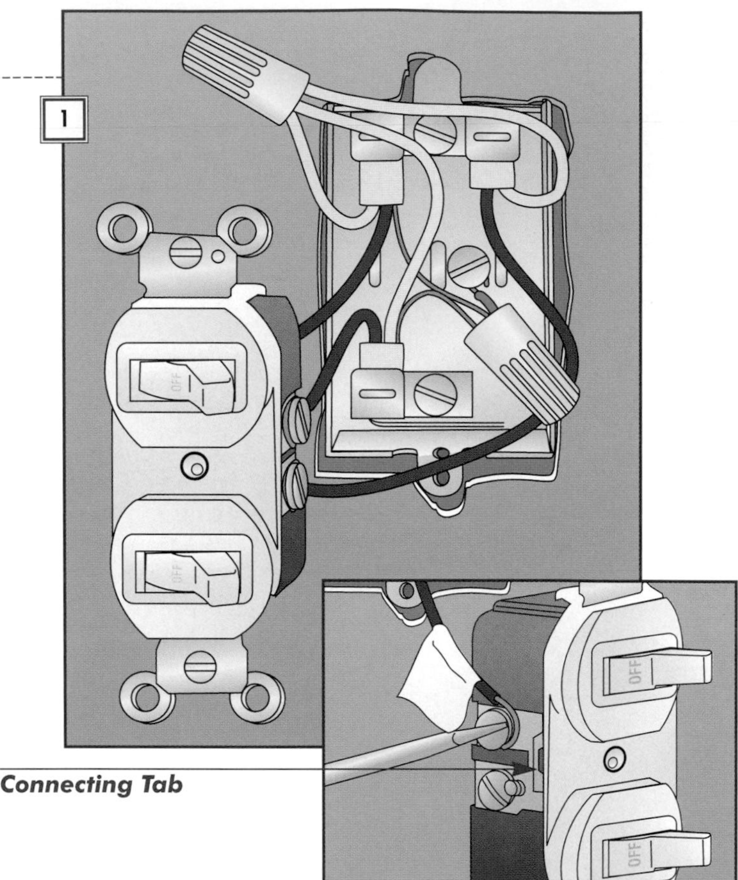


## 2. Testing for Continuity

- Disconnect the switch.

- Set a multitester to RX1. With the upper toggle in the OFF position, touch the probes to the upper terminals *(right).*

- Toggle the switch. If the switch is good, the meter will indicate continuity in the ON position but not when the switch is OFF.

- If the first switch passes, move the probes to the lower terminals and retest.

- If either switch fails, replace the combination switch, reattaching the wires as they had been connected to the old switch.

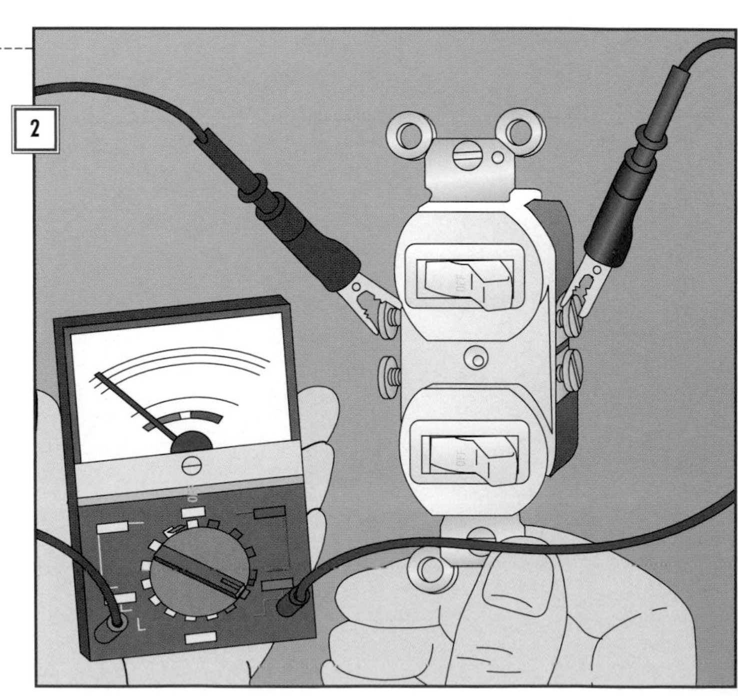

# Troubleshooting a Switch-Outlet

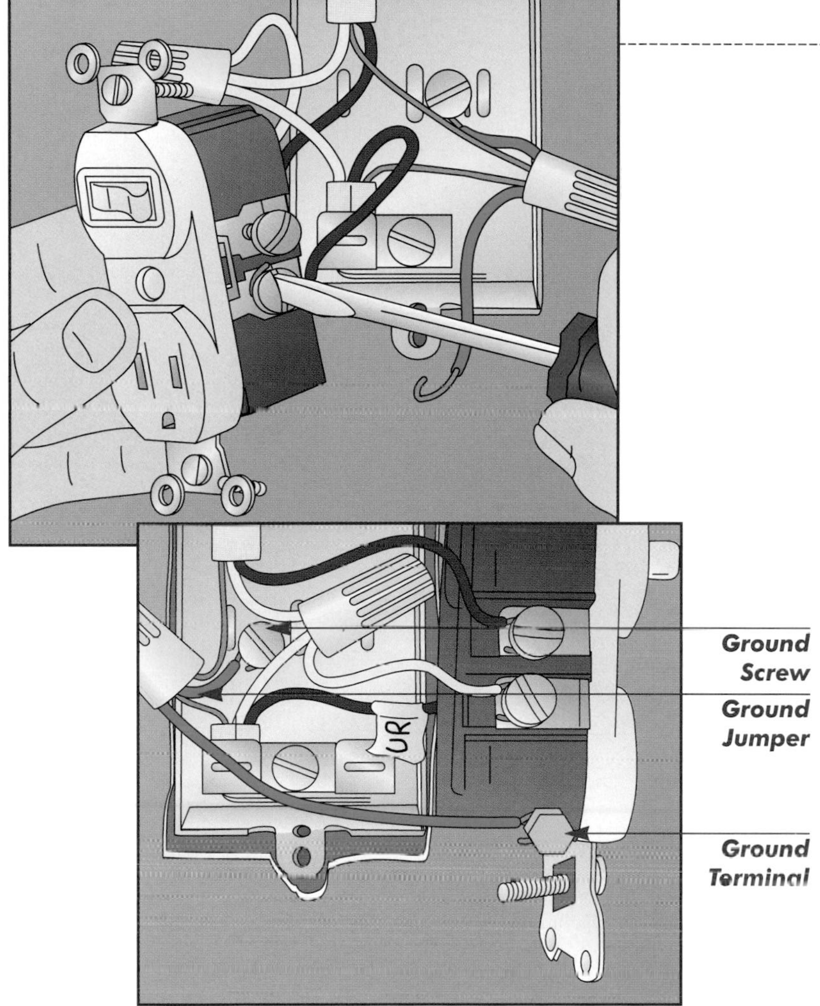


## When switch and outlet are independent

A switch-outlet combines a toggle switch with an outlet. The switch and outlet may work independently, or the switch may control the outlet. Shown here is the wiring for independent operation. (For troubleshooting a switch-outlet in which the switch controls the outlet, see page 80.)

- Shut off power to the circuit at the service panel *(page 10).*

- Remove the cover plate and test all connections for voltage *(page 68),* and free the switch from the box.

- Before loosening the terminals with a screwdriver *(upper left),* tag the black wire that leads to the brass terminal on the side of the switch with the connecting tab.

- Check and clean or repair the connections as necessary *(page 69).*

- Reconnect the wires, hooking the untagged black wire to the upper terminal on the opposite side of the switch, and the white wire to the lower terminal there *(lower left).* Attach the tagged black wire to the brass terminal on the side of the switch with the connecting tab. Reattach the ground connections as shown. Then refasten the switch to the box.

- Restore power to the circuit at the service panel. Flip on the switch to see whether the load that it controls gets power. Plug a lamp into the outlet to see if it works. If one of these tests fails, turn off the power and test the switch for continuity *(page 80).*

### When the switch controls the outlet

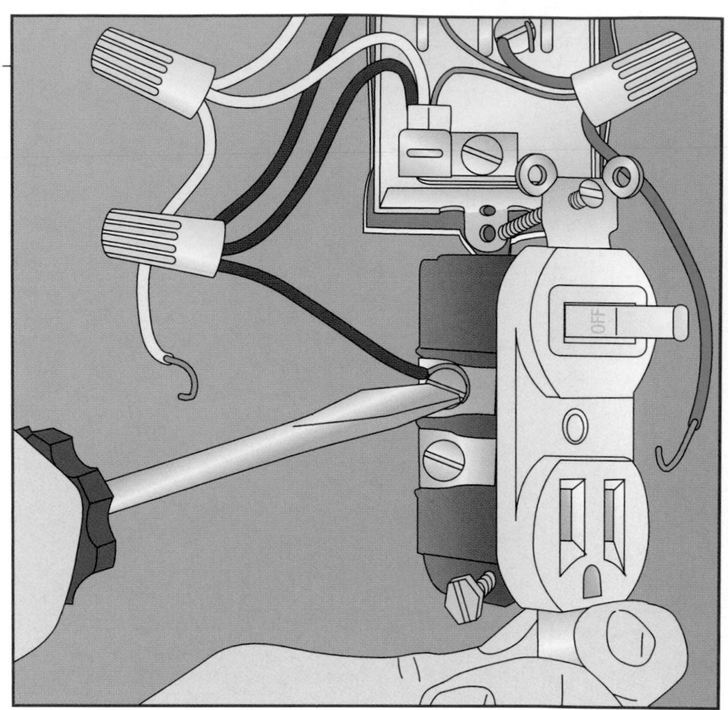


This illustration shows the typical wiring for a combination switch-outlet where the switch controls the outlet. Follow the troubleshooting procedures on page 79; then, if necessary, perform continuity tests as described below. To reconnect or replace the unit, proceed as follows:

- Connect the black jumper wire to the upper terminal on the side with no connecting tab *(right).*
- Fasten the white jumper around the lower terminal on the same side.
- Attach the ground jumper to the ground terminal.
- Fasten the switch-outlet into the box with the mounting screws, and turn on the power at the service panel.

### Testing for continuity

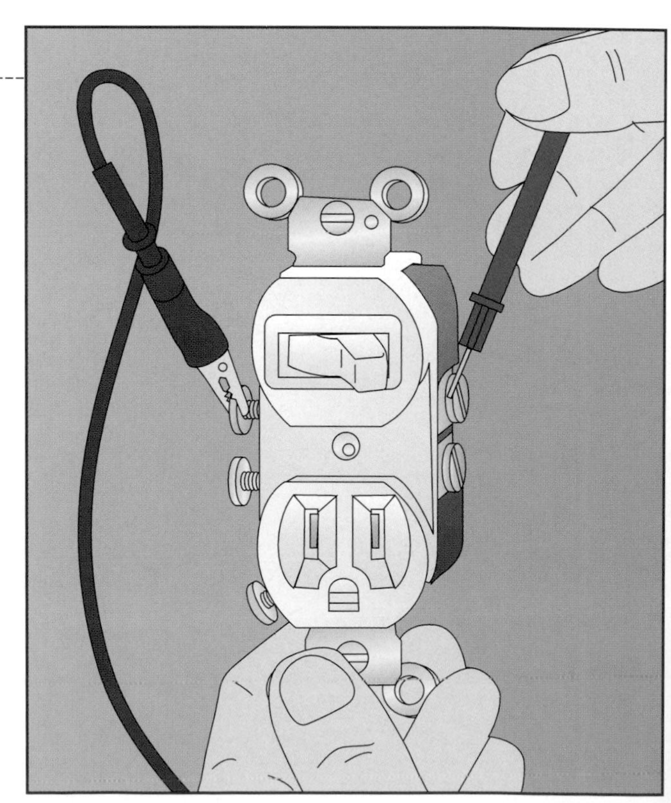

- After turning off the power and disconnecting the unit, set a multitester to RX1.
- Touch the probes to the two upper terminals, and toggle the switch. If the switch is good, the meter will indicate continuity in the ON position but not in the OFF position *(page 20).*
- If the unit fails the test, replace it, reattaching the wires as you found them.

# GROUNDING SWITCHES AND SWITCH BOXES

Grounding is an important safety precaution. If the cable in the box has a ground wire, any of the wiring plans shown here will result in good grounding.

**Metal Box with One Cable:** When the switch has no ground terminal, secure the cable ground wire to the screw at the back of the box. The screws holding the switch to the box complete the grounding circuit.

For a switch with a ground terminal, use a jumper and a wire cap to connect it to the cable ground wire and to a jumper from the ground screw in the box.

**Metal Box with Two Cables:** For a switch without a ground terminal, attach a jumper to the ground screw at the back of the box, twist it together with the bare ground wires from the two cables, and secure them with a wire cap. If the switch has a ground terminal, attach a jumper wire and join it with both cable ground wires and a jumper to the box.

**Plastic Boxes:** Where the switch has a ground terminal and the box doesn't, secure a jumper to the cable ground wire and fasten it to the switch ground terminal. If only the box has a ground terminal, fasten the cable ground wire to it. Where two cables enter a plastic box, splice ground wires from both to a jumper and connect it to the ground terminal on the box or the switch.

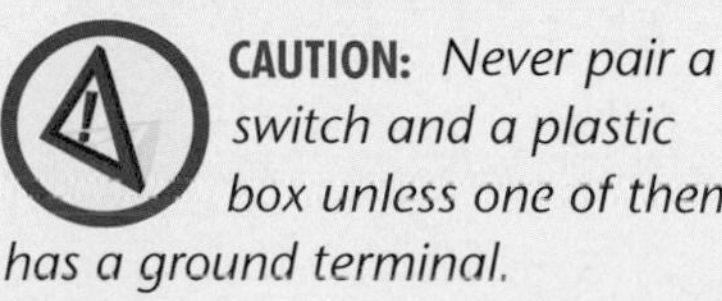

**CAUTION:** *Never pair a switch and a plastic box unless one of them has a ground terminal.*

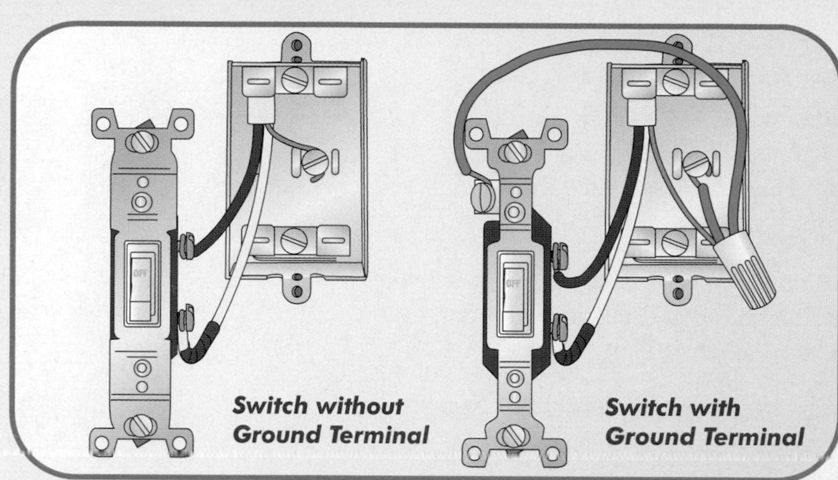


***METAL BOX WITH 1 CABLE***

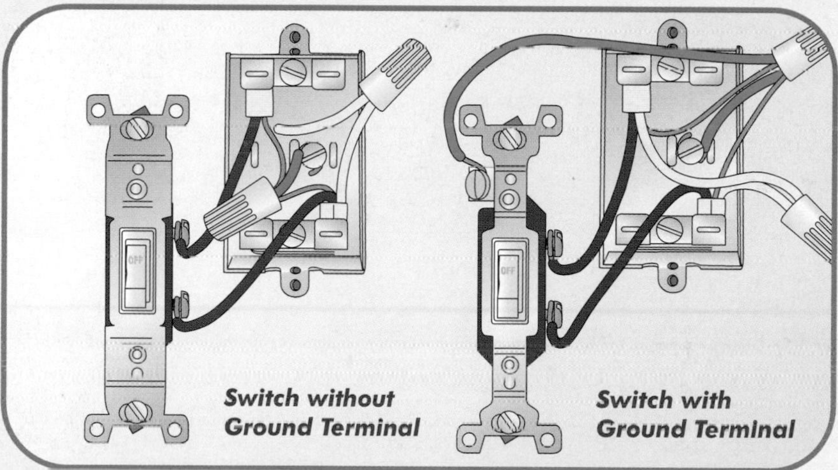


***METAL BOX WITH 2 CABLES***

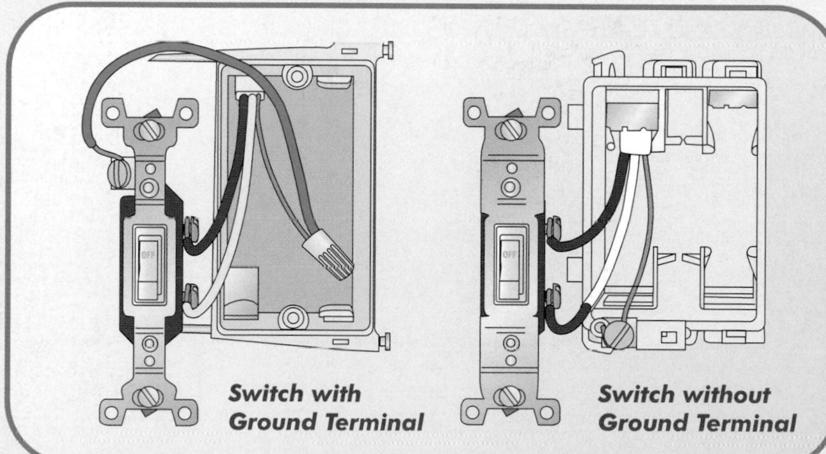


***PLASTIC BOX WITH 1 CABLE***

# Servicing a Faulty Pilot-Lamp Switch

### Tagging the hot-feed wire

A pilot-lamp switch is often used to control a light that can't be seen from the switch. A red bulb below the switch indicates when the light is on.

• Turn off power to the circuit at the service panel; then remove the cover plate and test for voltage *(page 68).*

• Before disconnecting the switch *(page 70),* tag the black wire attached to the brass terminal.

• Check and clean the connections as necessary *(page 69).* Reconnect the tagged wire to the brass terminal. Attach the white wire that you took off the switch to the terminal below the tagged black wire *(right).* Connect the other black wire to the upper terminal on the opposite side.

• Fasten the switch to the box with the mounting screws, turn on the power, and flip the switch to the ON position. If either the light or the red indicator fails, test the switch for continuity.

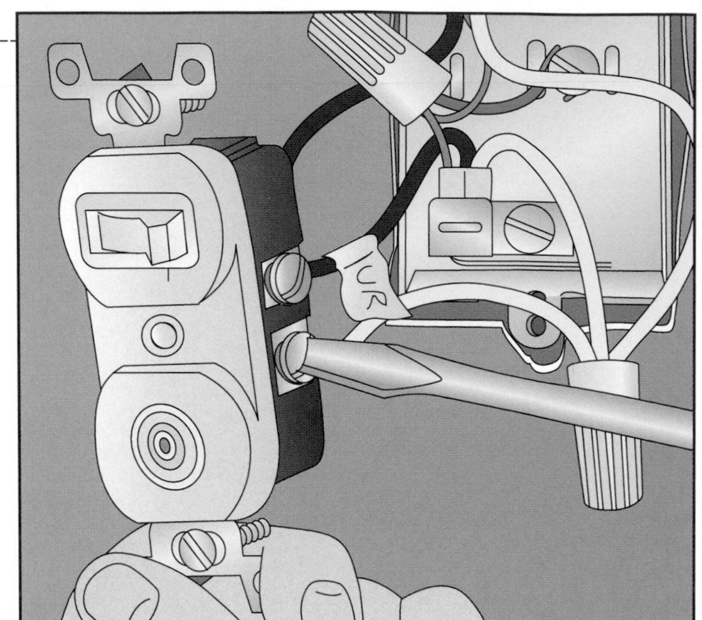

### Testing for continuity

• Turn off the power and disconnect the switch.

• Set a multitester to RX1. Flip the switch to the ON position, and touch or clip the probes to the switch's upper terminals. In a good switch, the meter will indicate continuity in the ON position but not in the OFF position *(page 20).*

• If the switch fails the test, replace it, attaching the wires as you found them.

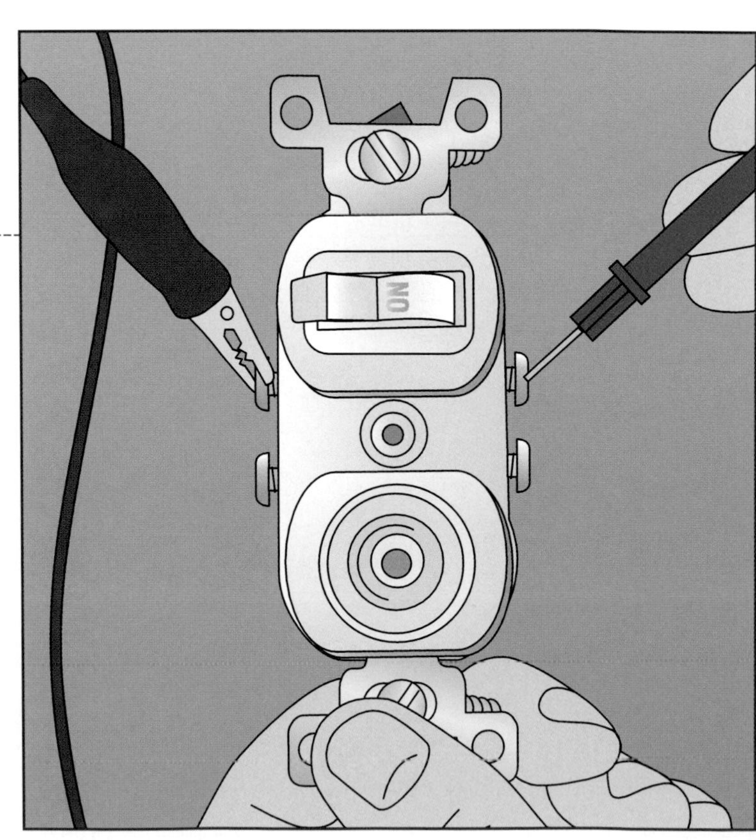

# Testing and Replacing a Timer Switch

## 1. Dismounting the Timer

A timer switch turns a fixture on and off automatically at intervals set by the user.

- Turn off power to the circuit at the service panel, then unscrew the timer from the wall *(left).*

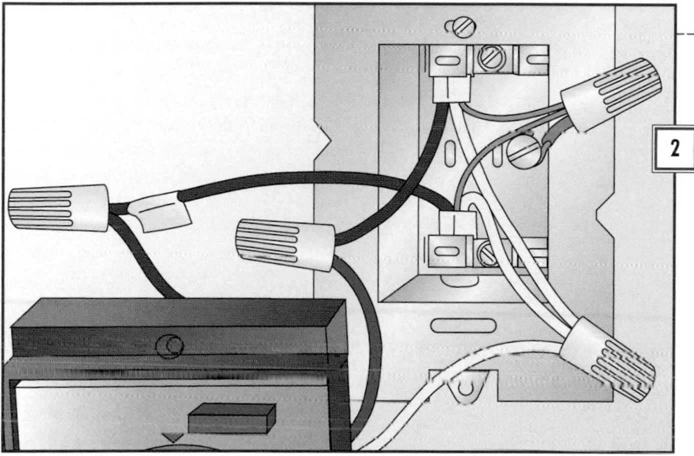


## 2. Tagging the Hot-Feed Wire

- Tag the black wire connected to the timer's black lead with masking tape *(left).*
- Carefully remove the wire caps without touching the metal, and test for voltage at each connection *(page 68).*
- Check, clean, and service the connections as necessary *(page 69).* If there are no apparent loose or burned connections, test the switch for continuity *(Step 3).*

## 3. Testing for Continuity

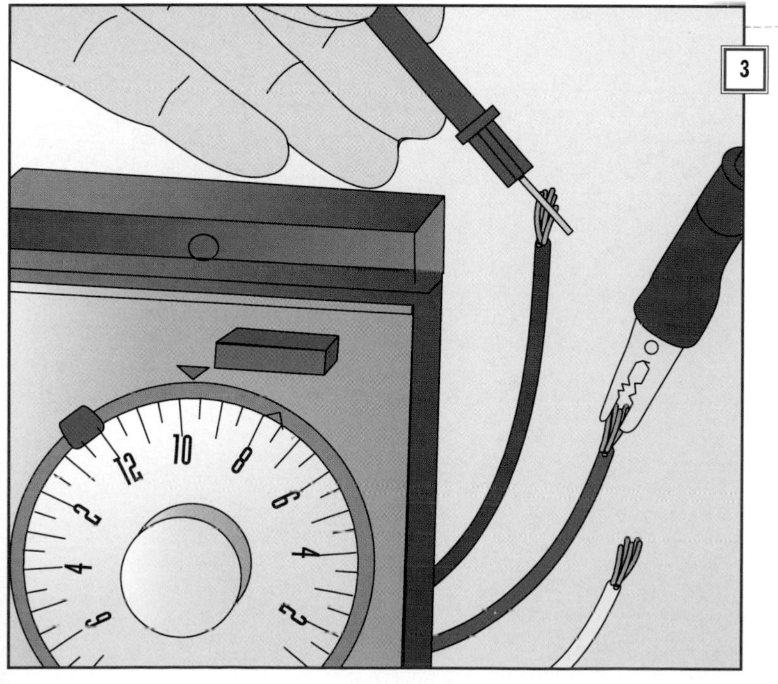


- Set a multitester to RX1. Place one probe on the black lead and the other on the red lead, as shown at left.
- Press the manual control button to turn on the switch. The meter should indicate continuity *(page 20).* If not, replace the switch.
- To install the new switch, connect the black switch lead to the tagged black wire.
- Use a wire cap to join the red switch lead and the untagged black wire.
- Connect the white switch lead to the pigtail of white wires in the box.

Chapter 5

# FIX IT: Wall Outlets

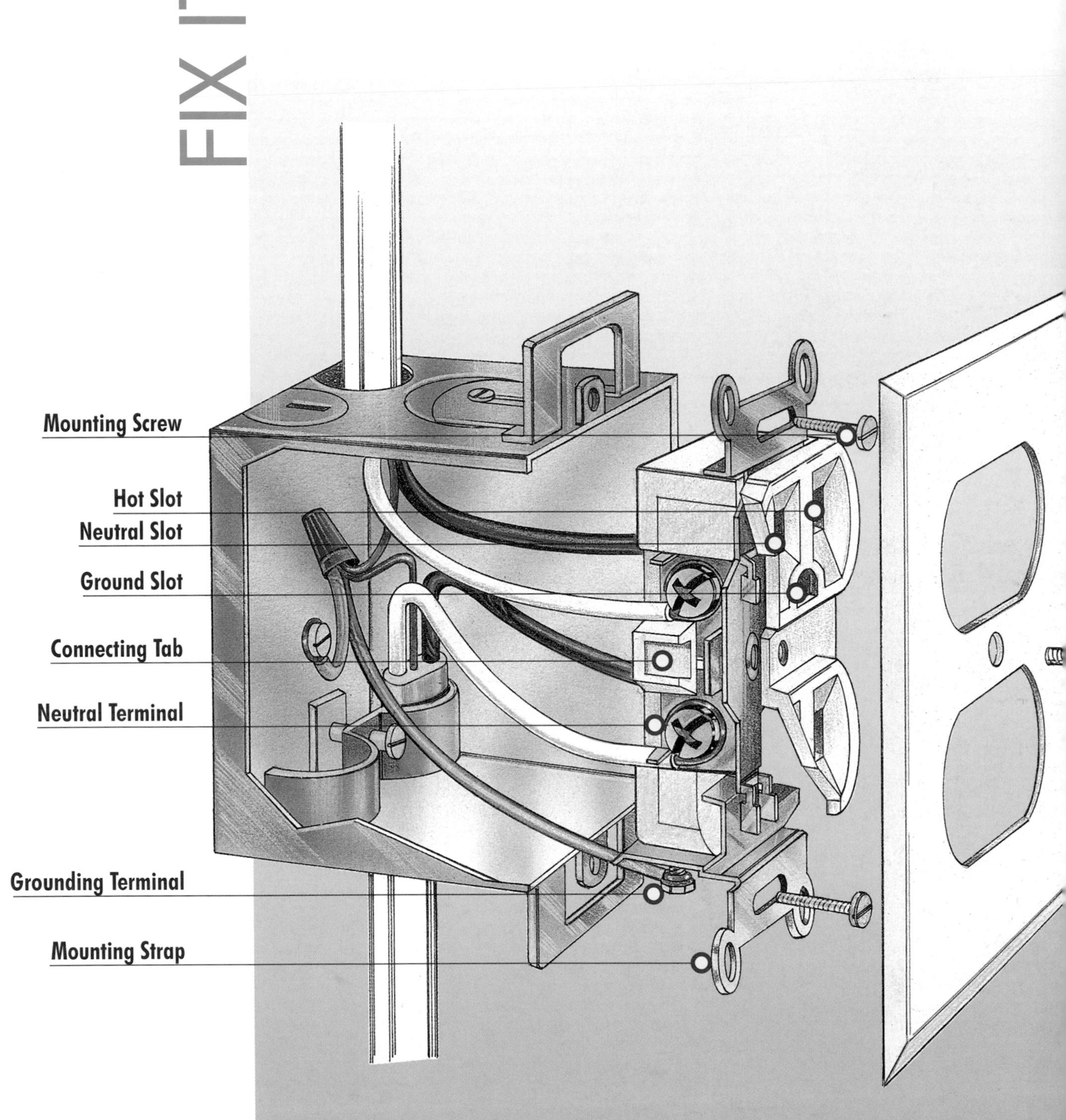

# Chapter 5

## How They Work

The illustration at left shows a modern 120-volt, 15-amp duplex (two-receptacle) wall outlet. Each receptacle has three openings. One leads to the black (hot) wire, always connected to a brass terminal screw, that brings power from the service panel. A shorter slot beside it leads to the white, or neutral, wire (silver terminal). The third, round slot leads to a grounding terminal.

This arrangement, mimicked by the prongs of plugs, assures that an appliance can be plugged into the outlet only one way. Doing so reduces the likelihood of electrical shock by preventing the hot side of the circuit from being connected to easily accessible parts of the appliance.

A duplex outlet also has a connector tab that links the two receptacles electrically and assigns both to the same circuit. Breaking off the tab permits each receptacle to be on a different circuit or to have one receptacle but not the other controlled by a wall switch.

## Contents

# Troubleshooting

| Problem | Solution |
|---|---|
| • Plugged-in appliance does not work | Check appliance • Replace fuse or reset circuit breaker **10** • Check outlet connections **89** • |
| • Fuse blows or breaker trips when appliance is turned on | Move appliance to another circuit • Service or replace appliance • Look for a short in the wiring • |
| • Appliance plug keeps falling out | Replace plug **26–28** • Replace outlet **90** • |
| • Appliance runs intermittently or lamp flickers | Call technician to check appliance • Repair lamp **18–36** • Check outlet connections **89** • |
| • Sparks or mild shock when plugging in appliance | Turn appliance off before plugging it in • Hold plug by insulation when plugging it in • Service or replace appliance • Check outlet connections **89** • |
| • Switch-controlled outlet does not deliver power | Test wall switch and replace **72** • Check outlet connections **89** • |
| • Only one socket in split-circuit outlet works | Check connections **89** • Replace outlet **95** • |
| • GFCI trips repeatedly | Service or replace appliances on the circuit • Check outlet connections **89** • Have an electrician locate ground fault • |
| • GFCI does not trip off when test-button is pushed | Replace outlet **94** • |
| • 2-slot outlet is connected to a grounded circuit | Replace with 3-slot outlet **90–93** • Install grounding adapter **91** • |
| • Outlets are not connected to a grounded circuit | Install a GFCI-equipped outlet in each ungrounded circuit **94** • |
| • 240-volt outlet does not work | Replace fuse or reset circuit breaker **10** • Check the connections and/or replace outlet **97–98** • |

# Before You Start

Outlets rarely break, but they sometimes wear out. Replacing an outlet is usually a simple matter of transferring wires from the old receptacle to a new one.

### Normal Wear and Tear

The statement above is true of not only standard, duplex outlets, but also the other types of outlets you may find in your home. An outlet equipped with a ground-fault circuit interrupter (GFCI) turns off power instantly when it detects potentially dangerous current leakage (*page 94*). And there are 240-volt outlets for larger appliances (*page 97*), which are supplied by two 120-volt hot conductors.

The most common problem with outlets that have been in service for many years is worn internal contacts—plugs feel loose when you insert them in the outlet. Replace the outlet if its contacts are worn (*page 90*).

Wires connected to outlets can work themselves loose with ordinary expansion and contraction or from vibration. This can cause sparking and a potential fire hazard. Always tighten terminal screws firmly on wires and avoid using the push-in terminals found on the back of many outlets.

### Before You Start Tips:

- A white wire marked with black tape or paint serves as a hot wire; connect it to a brass terminal.
- If your house wiring has no grounding, you can provide this protection against fire and shock by installing a GFCI outlet on each circuit as the first outlet that the cable encounters after leaving the service panel.

Multitester
Receptacle analyzer
Screwdriver
Long-nose pliers
Lineman's pliers
Wire stripper or multipurpose tool

Wire
Wire caps
Electrical tape
Replacement outlets

When working on an outlet, never touch any terminals or wire ends until you have confirmed by a voltage test that you have disconnected power to the outlet.

# Duplex Outlets

## 1. Removing the cover plate

- Turn off the power to the outlet by removing the fuse or tripping the circuit breaker at the service panel *(page 10)*.
- Remove the cover plate screw *(right)*.
- Set the cover plate aside. If it's stuck to the wall, cut neatly around its edges with a utility knife.
- Tape the screw to the cover plate to avoid losing it.

## 2. Testing a 120-volt outlet for voltage

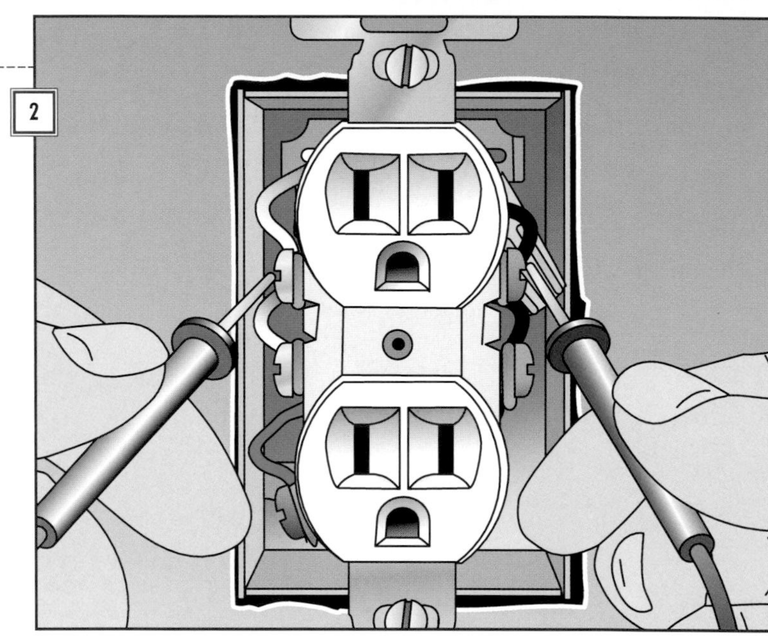


- To confirm that the power is off, set a multitester to 250 volts in the AC-voltage range *(page 20)*.
- Touch one probe to a brass terminal where a wire is attached, and the other to a silver terminal where a wire is connected *(right)*. Then touch one probe to the the grounding terminal, and the other to the brass and silver terminals in succession. The multitester should register 0 voltage.
- Repeat the test on the second pair of brass and silver terminals if wires are attached to them.
- If any test indicates that voltage is present, return to the service panel and turn off power to the correct circuit.

**CAUTION:** *Since there is a possibility that current may be flowing to the outlet, be sure to hold the multitester probes by their insulated grips.*

## 3. Freeing the outlet from the box

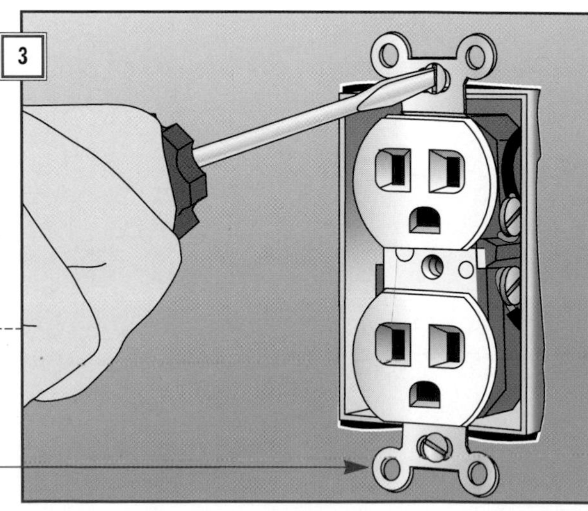


- Loosen the mounting screws *(right)*.
- Grasp the mounting strap, and pull the outlet out of the box.

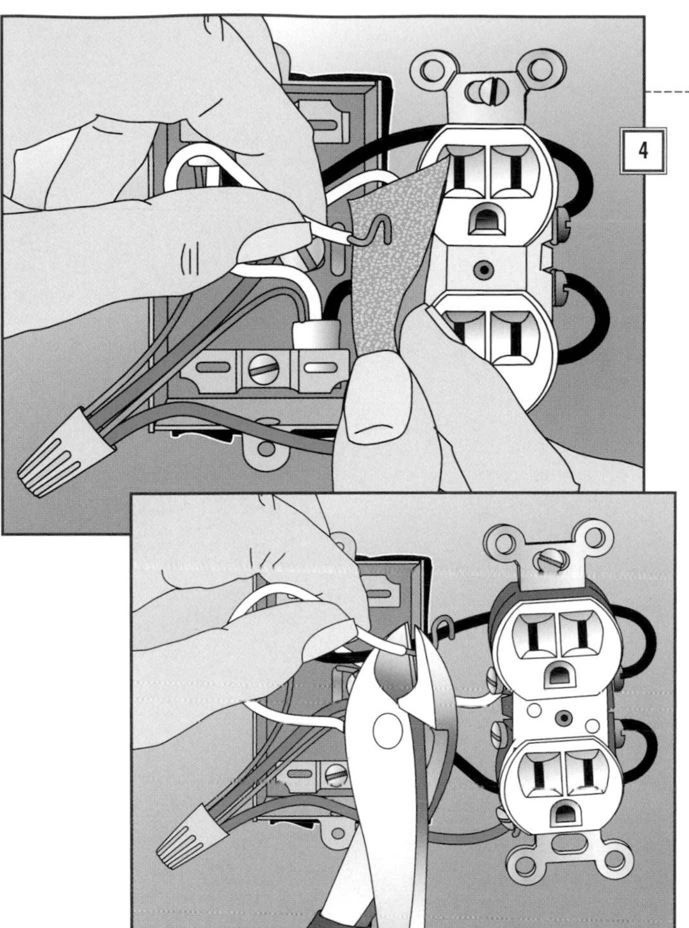


## 4. Checking and Repairing the Connections

Dirty or loose connections can produce sparks, shocks, or outlet failure. Inspect connections carefully.

- If a connection is loose, reshape the loop at the wire end with long-nose pliers *(page 69),* and tighten the terminal screw securely against the wire.
- For a dirty or corroded connection, disconnect the wire, then burnish it with fine sandpaper *(left, top).*
- Clip off the end of a discolored wire with diagonal-cutting pliers *(left, bottom),* then use wire strippers to remove 3/4 inch of insulation. Shape the wire end into a loop, and curl it around its terminal.
- If outlet terminals appear blackened or burned, replace the outlet *(page 90).*
- Fasten the outlet in the box, put on the cover plate, and restore power to the circuit. Plug in a lamp to see if the outlet is working; if not, replace the outlet.

## WHEN A WIRE IS TOO SHORT

Sometimes a damaged wire may be too short for you to form a loop in the end and reconnect it directly to its terminal *(right, top).* The solution is to attach a jumper wire to the terminal and splice it to the broken wire with a wire cap *(right, bottom).*

To make the jumper, cut a 4- to 6-inch length of wire having the same gauge as the wire in the box. With a wire stripper or multipurpose tool, remove 3/4 inch of insulation from each end. Form a loop at one end with long-nose pliers *(page 69).* Strip about the same amount of insulation off the end of the short wire remaining in the box.

With lineman's pliers, twist it together with the free end of the jumper, securing the connection with a wire cap *(page 73).* Hook the loop around the terminal screw and tighten it.

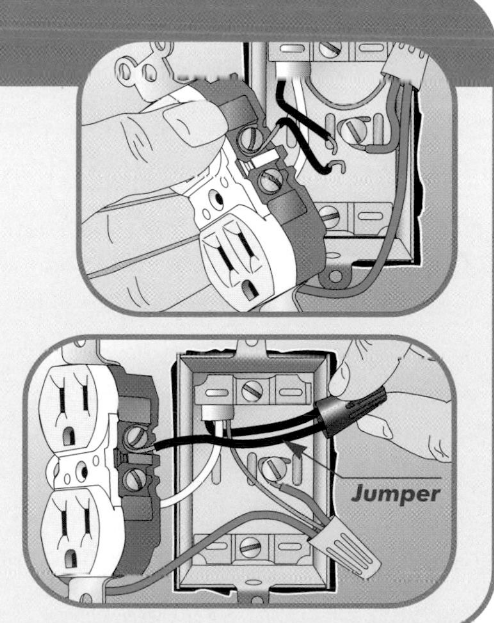

### 5. Replacing the outlet

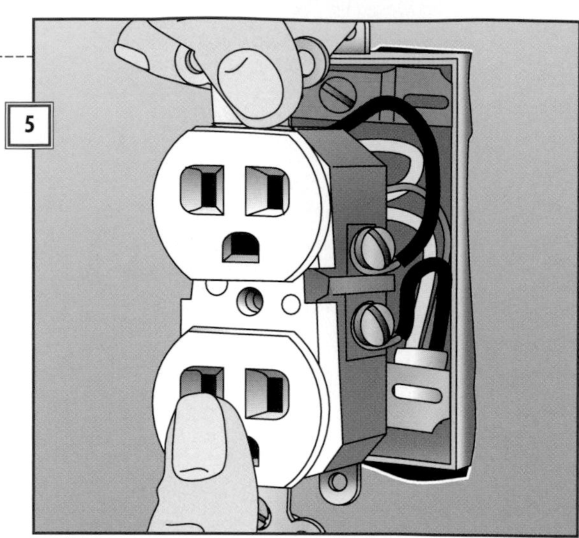


• Connect the wires as you found them connected to the old outlet: black wires to brass terminals, white wires to silver terminals, and the grounding jumper to the green grounding terminal. Tighten the terminal screws securely.

• Gently fold the wires into the box and set the outlet in place *(right).* Screw the mounting strap to the box, making sure the outlet is vertical.

• Restore the cover plate and turn on the power.

## Upgrading a 2-Slot Outlet

### 1. Testing a 2-slot outlet for ground

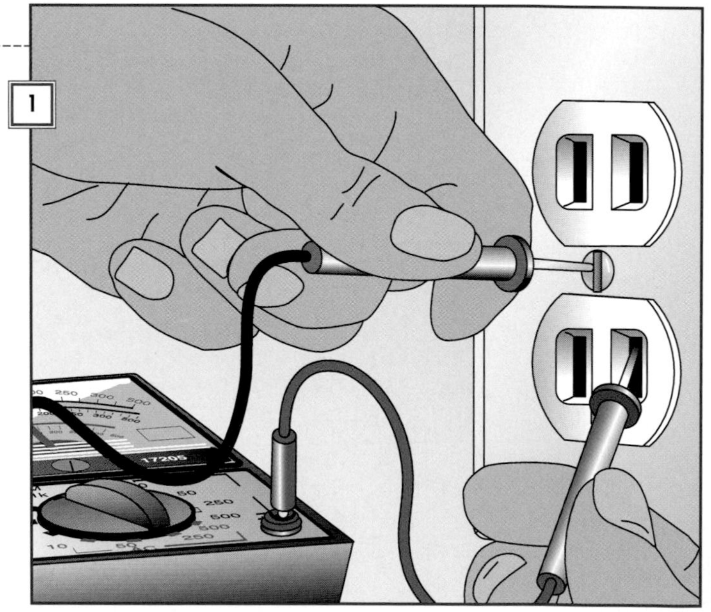


You can replace a 2-slot outlet with a 3-slot grounded outlet if the outlet box is grounded.

• Set a multitester to 250 volts in the AC-voltage range *(page 20).* Place one probe on the cover plate screw (scrape off paint as needed to uncover bare metal) and the other into one slot, then the other *(right).* The meter should indicate full voltage when the probe is in one slot, but not the other.

**CAUTION:** *This is a live voltage test. Be sure to hold the multitester probes by their insulated handles.*

• A voltage reading of 0 at both slots means the box isn't grounded. Have an electrician extend a grounded circuit to it, or install a GFCI in the box *(page 94).*

• If the test shows that the outlet is grounded, turn off power to the circuit at the service panel. Test for voltage *(page 88)* to make sure power is off, then proceed to Step 2 *(page 92).*

## HOW TO GROUND AN OUTLET

You can readily check whether an outlet is correctly grounded by checking the wiring. Be sure to turn off power to the circuit at the service panel before pulling the outlet from the box.

In a metal box *(right, top),* a properly grounded outlet has two jumper wires (green) joined to the bare wires in the box by a wire cap. One jumper wire goes to the grounding screw at the back of the box; the other, to the outlet's grounding terminal. In a plastic box *(right, bottom),* the wiring is identical, except that there is no jumper to the box.

When you find grounding that differs from these examples, use the techniques shown on pages 92 and 93 to correct it.

If a 2-slot outlet has been installed in a grounded box *(page 90),* you can fit an adapter for 3-prong plugs *(below).* Remove the mounting screw on the cover plate, plug in the adapter, then reinstall the screw. Perform a voltage test *(page 90)* to confirm that the screw is still grounded.

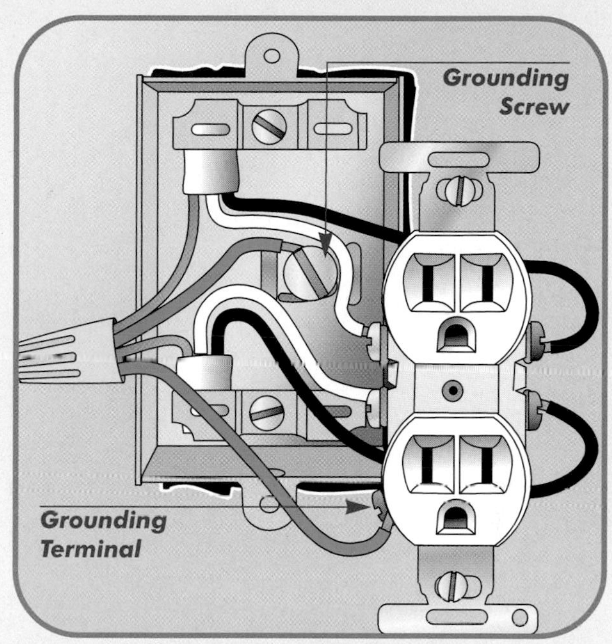


GROUNDED OUTLET IN A METAL BOX

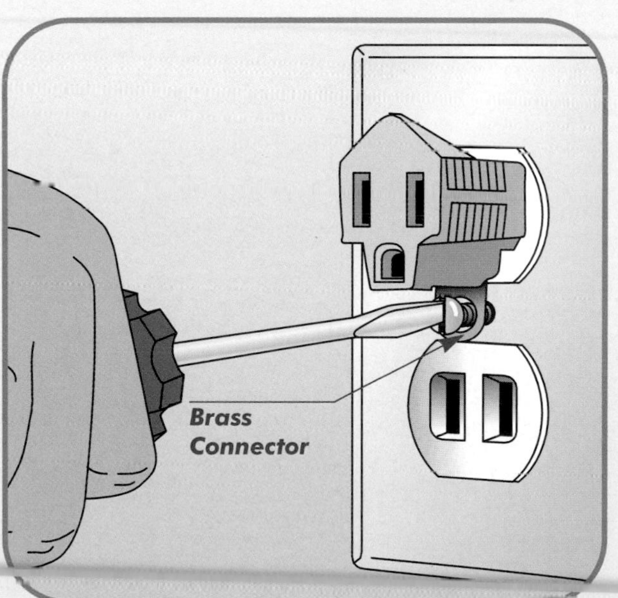


A GROUNDING ADAPTER

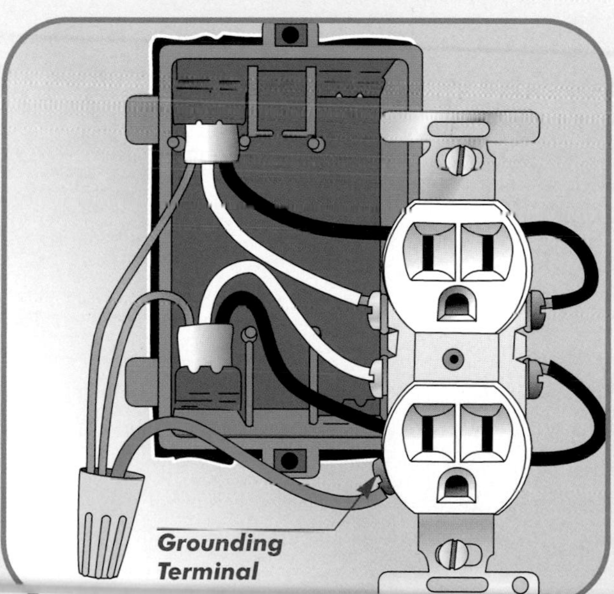


GROUNDED OUTLET IN A PLASTIC BOX

### 2. Creating a grounding pigtail

The term *pigtail* refers to any short wire—generally 6 inches or less—but may also refer to an assembly that includes one or more short wires. It is often more convenient to splice short wires into a pigtail before making connections within the box.

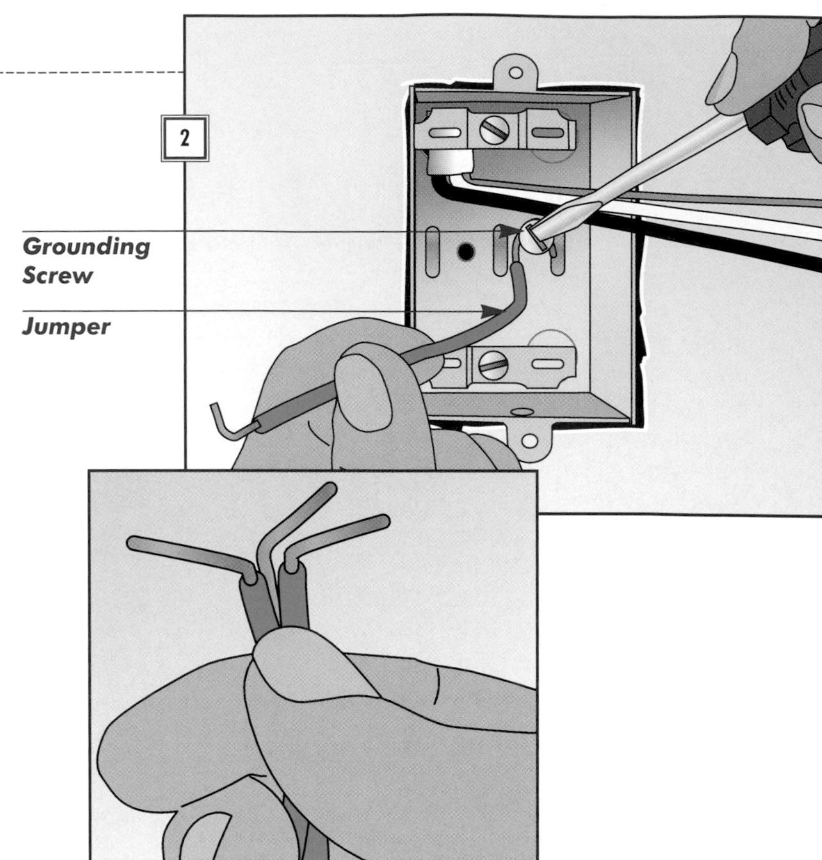


• Cut a length of wire—the same gauge as the wire in the box—4 to 6 inches long. Strip about 3/4 inch of insulation from each end *(page 110).* Form a loop at one end *(page 69),* and attach it to the grounding screw at the back of the box *(right).* Then bend the free end of the jumper to a right angle.

• Prepare the second jumper. Strip insulation from both ends. Form a loop at one end, and bend the other to a right angle.

• Bend the end of the ground wire from the cable entering the box to a right angle.

• Gather the ends of the wires with right-angle bends in one hand *(inset).* With lineman's pliers, grip the wire ends and twist them together in a clockwise direction until the turns are tight and uniform along the entire length of the bare wire. Straighten the splice; then clip 1/4 inch off the end of the twisted wires, and secure the connection with a wire cap *(page 73).*

### 3. Connecting the grounded outlet

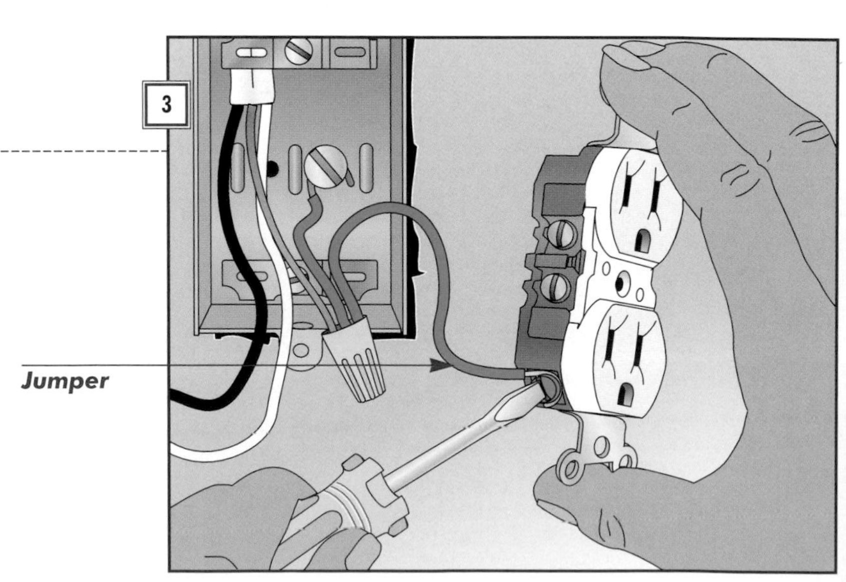


• Hook the free jumper from the 3-wire grounding pigtail around the green terminal on the outlet, and tighten the terminal screw securely *(right).*

• Hook the black wire around a brass terminal and the white wire around a silver terminal. Tighten the terminal screws.

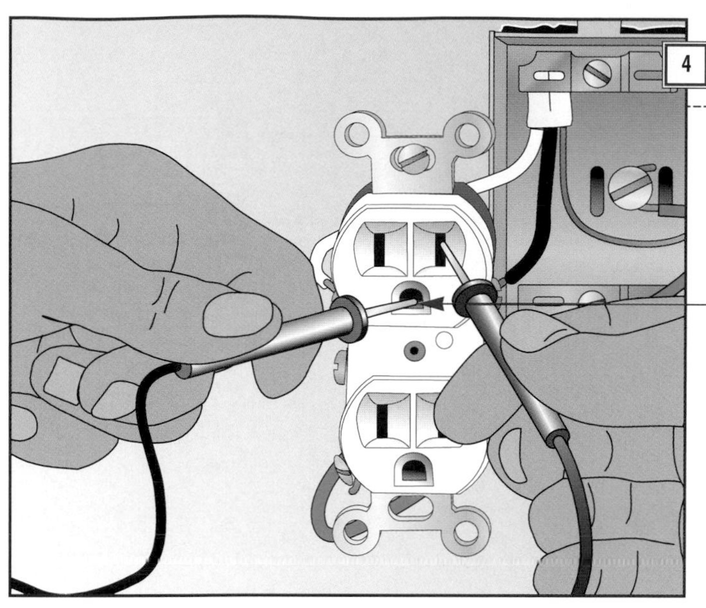


## 4. CHECKING A 3-SLOT OUTLET FOR GROUND

- Restore power to the circuit at the service panel.
- With the multitester set to 250 volts in the AC-voltage range, insert one probe into the round grounding slot and the other probe into the other two slots in turn.

**CAUTION:** *This is a live voltage test. Be sure to hold the multitester probes by their insulated handles.*

- The meter should indicate voltage when the second probe is in the short slot *(left)* but not in the taller slot.
- If the outlet fails the test, turn off the power and check the connections.

## A MULTITESTER ALTERNATIVE

Widely available at home centers, hardware stores, and electrical-supply houses, receptacle analyzers *(right)* speed and simplify outlet tests. Simply plug one into a receptacle you wish to check. Three display lights indicate whether there is current to the receptacle, whether the hot and neutral wires are reversed, and whether the outlet is properly grounded. Always test both receptacles in an outlet. If the analyzer reveals a wiring fault, check the outlet connections or replace the outlet.

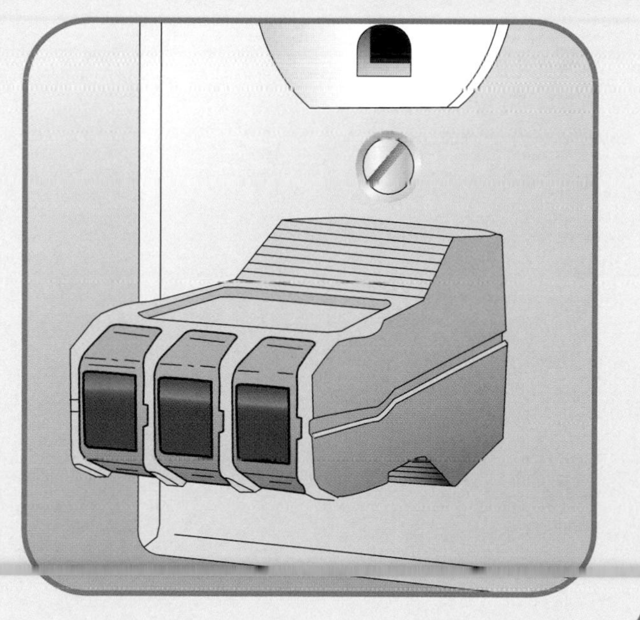

# GFCI Outlets

## 1. Testing the outlet

For safety, a GFCI-equipped outlet shuts off power when it senses a difference in amperage between the hot and neutral wires caused by current leaking to ground or another conductor that does not carry current.

- Press the test button *(right)* once a month; if the reset button fails to pop out, replace the outlet.

- If a GFCI outlet trips repeatedly, call an electrician to find the fault on the circuit.

## 2. Replacing the outlet

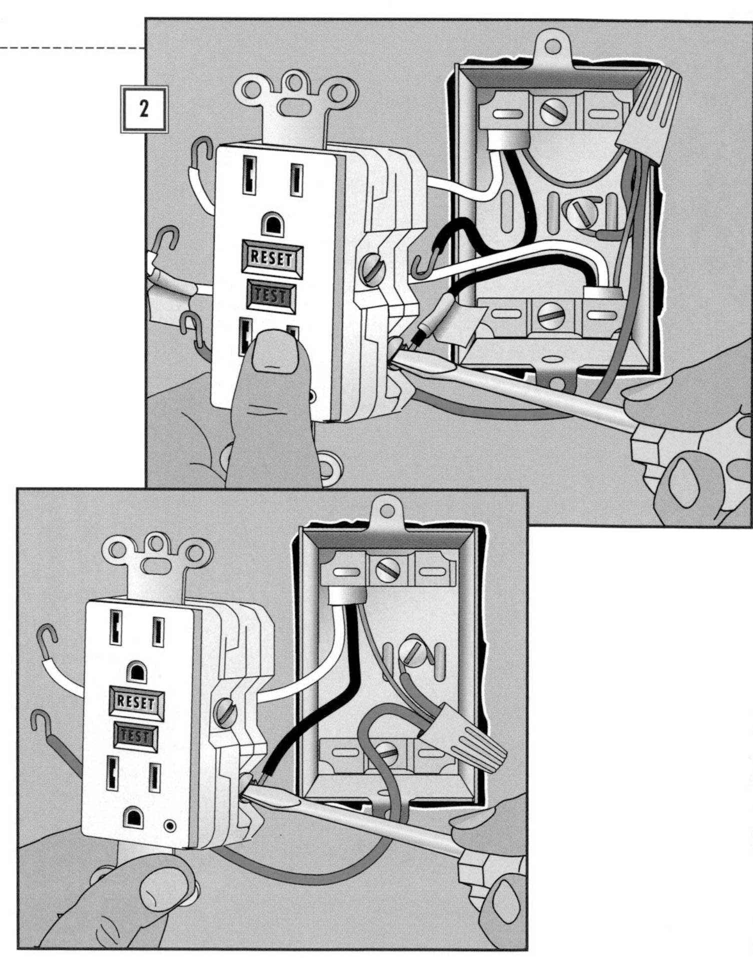


- Turn off power to the circuit at the service panel *(page 10)*. Remove the cover plate and test for voltage *(page 88)*. If voltage is not 0, return to the service panel and turn off another circuit.

- Pull the outlet from the box. If two cables enter the box, tag the black and white wires connected to the terminals marked LINE on the back of the outlet, then disconnect all the wires *(right)*. (You need not tag wires in a 1-cable box.)

- When installing a GFCI outlet in a 2-cable box, connect the tagged black wire to the brass terminal marked LINE, and the tagged white wire to the corresponding silver terminal. Connect the remaining black and white wires to the other brass and silver terminals (LOAD).

- In a 1-cable box *(inset)*, the black wire goes to the brass terminal marked LINE, and the white wire goes to the corresponding silver terminal.

- Connect the grounding wire to the green terminal.

- Reinstall the outlet, turn on the power, and press RESET.

# Split-Circuit Outlets

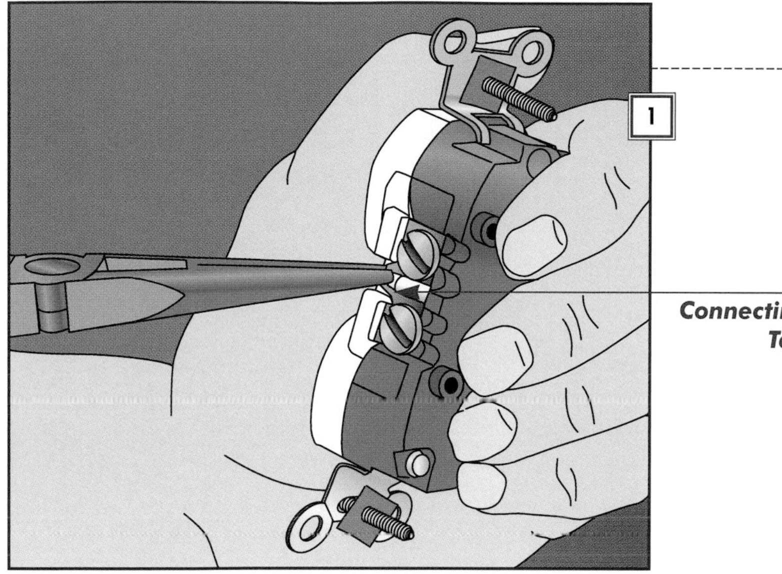


### 1. Removing the Hot-Side Connecting Tab

Usually, a single circuit serves both receptacles in an outlet via separate connecting tabs that link both brass (hot) terminals and both silver (neutral) terminals. But when a 240-volt circuit is involved, the receptacles can operate independently on 120 volts, an arrangement called a split circuit. You can recognize a split-circuit outlet by the absence of a connecting tab linking the brass terminals.

- Turn off power to the circuit at the main service panel or at an auxillary panel nearby *(pages 10–11).*

- Remove the outlet cover and test for voltage *(page 88)* to make sure that power to the circuit is off.

- Check and service the connections *(page 89).* If the outlet still works poorly after you've reinstalled it and restored power, buy a standard 120-volt replacement.

- Snap off the connecting tab between the two brass terminals by bending it back and forth with long-nose pliers *(above left).*

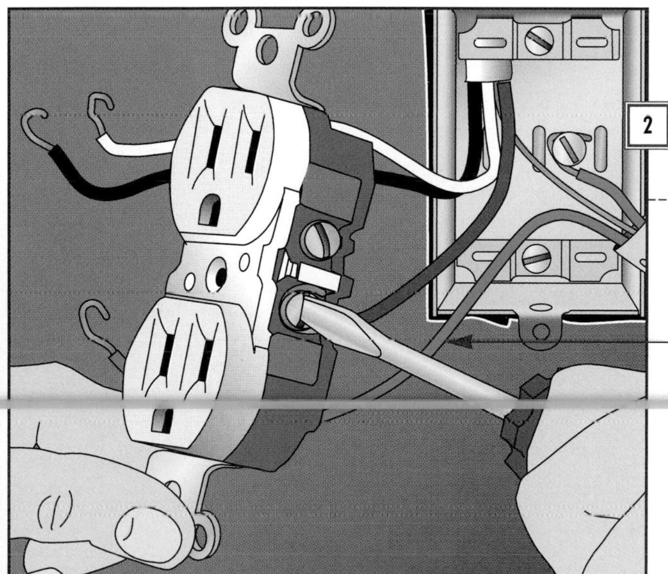


### 2. Installing the New Outlet

- Attach the red wire to one brass terminal on the outlet *(left),* and the black wire to the other brass terminal. Connect the white wire to either of the silver terminals, and the grounding jumper to the green grounding terminal.

- Gently fold the wires into the box.

- Screw the outlet into the box, making sure that the outlet is vertical; then install the cover plate and turn on the power.

# Switch-Controlled Outlets

## When the switch controls both receptacles

When a switch-controlled outlet isn't working, replace it only after first checking the switch *(page 68)* and the outlet connections *(page 89).*

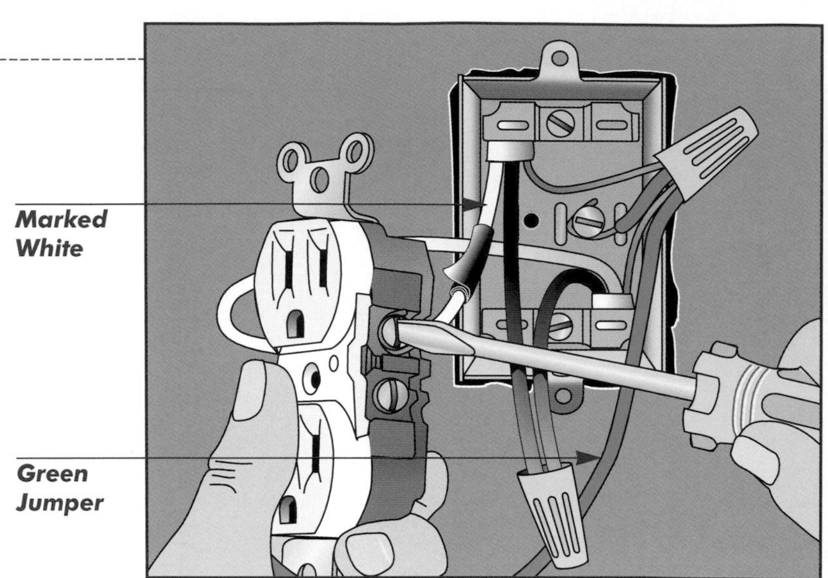


• Turn off power to the circuit at the service panel *(page 10).* Remove the outlet cover plate and test for voltage *(page 88).* If the power is off, pull the outlet from the box and detach the wires.

• If two cables enter the box, you are likely to find two white wires—one marked with tape or paint—attached to the outlet. Connect this wire to a brass terminal on the new outlet *(right).* Attach the other white wire to a silver terminal. Where there's only one cable, the black wire goes to a brass terminal, and the white wire to a silver one.

• Reconnect the green jumper to the outlet's green terminal.

• Fasten the outlet to the box, screw on the cover plate, and then restore power.

## When the switch controls one receptacle

• Break off the connecting tab between the outlet's brass terminals *(page 95).*

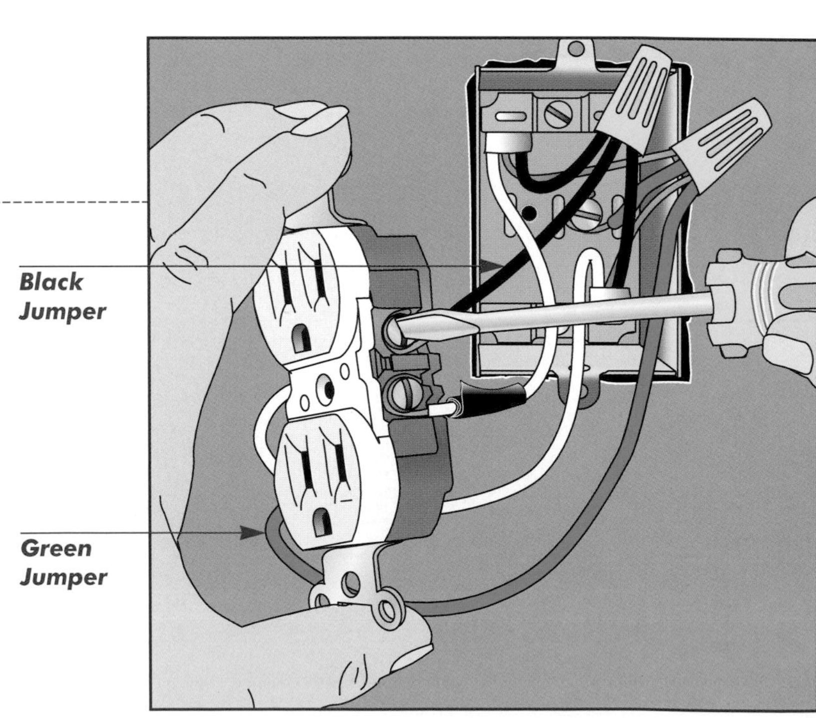


• For a 2-cable box *(right),* hook the marked white wire and the black jumper to the brass terminals. Connect the unmarked white wire to a silver terminal, and the grounding jumper to the green terminal.

• If there is one cable in the box, connect the black wire to one brass terminal, the red wire to the other brass terminal, and the white wire to a silver terminal. Attach the green jumper to the green terminal.

• Fasten the outlet to the box, screw on the cover plate, then restore power.

# 240-Volt Outlets

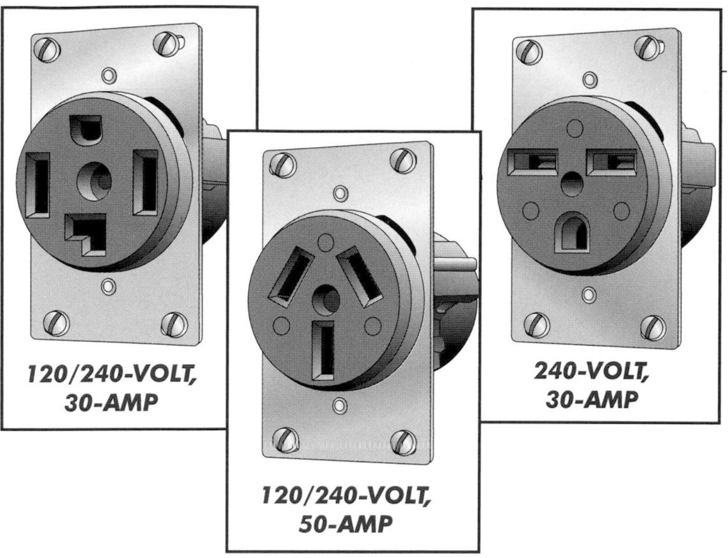


## Three Different Types

A 240-volt circuit has two hot wires; each feeds 120 volts of current to separate terminals of a 240-volt outlet. There are variations in the design and function of 240-volt outlets—the arrangement and shapes of the slots ensure that the plug of an appliance designed for one type of outlet can't be plugged into another type. A 120/240-volt, 30-amp outlet is used for appliances like dryers that require 240-volt current for their electric heating elements and 120-volts for their motors and timers. A 120/240-volt, 50-amp outlet may be used for kitchen ranges with 240-volt burners and 120-volt clocks and lights. A 240-volt, 30-amp outlet is often used with air conditioners, water heaters, and other large appliances.

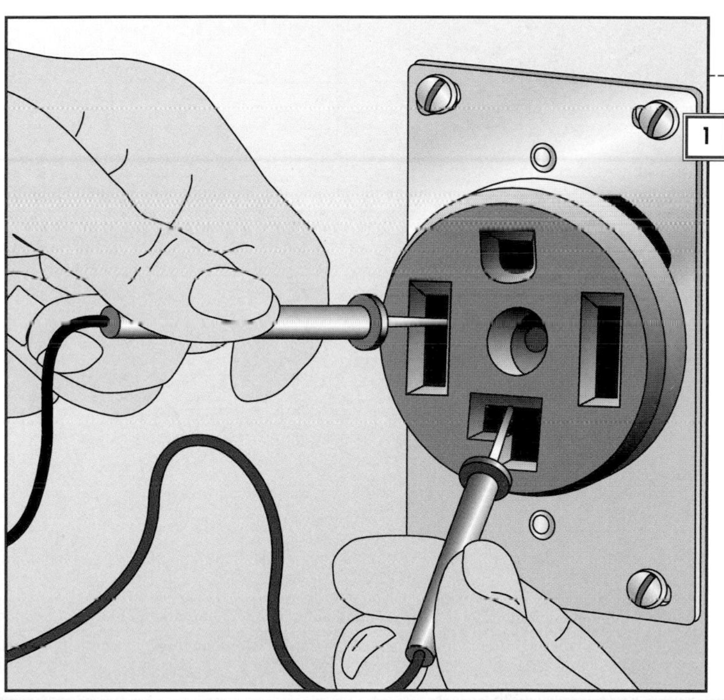


## 1. Checking for Voltage

If an appliance plugged into a 240-volt outlet seems to get no power, check the circuit-breaker or fuse panel first *(page 10).*

• If you can't solve the problem by resetting the circuit breaker or replacing the fuse, set a multitester to 250 volts in the AC-voltage range *(page 20).* Insert one probe into the lower (neutral) slot, and the other into each of the upper slots, in turn *(left).* The meter should indicate around 120 volts for each test. Then insert the probes in the two vertical slots; the meter should indicate around 240 volts.

**CAUTION:** *This is a live voltage test. Be sure to hold the multitester probes by their insulated handles.*

• If you obtain different results, the outlet connections need attention or you need a new outlet *(page 98).* Turn off power at the service panel, and retest for voltage to be sure that the circuit is off.

## 2. Checking the connections

This process is similar to checking the connections on a 120-volt duplex outlet *(page 89).* These illustrations show wiring for a 120/240-volt, 30-amp outlet *(right)* and a 240-volt, 30-amp outlet *(inset).*

- Take off the cover plate.
- Release the outlet from the box by removing the screws on the top and bottom of the mounting plate. Grasp the mounting plate, and pull the outlet from the box.
- Loosen the terminal setscrews on the back of the outlet, and pull out each wire in turn. Note that the stripped wire ends are straight, not formed into loops.
- If the connections are dirty or corroded, detach the wires and clean the wire ends with fine sandpaper. If necessary, clip and strip the wire ends *(page 89).*
- Push each wire into its terminal hole, making sure that no uninsulated wire is exposed. Tighten the setscrews to secure the connections and screw the outlet back into the box.

2

***Setscrew***

## 3. Testing the outlet

- Turn on the power and, with a multitester, check for voltage *(page 97)* to confirm that the outlet is getting power *(right).*

**CAUTION:** *This is a live voltage test. Be sure to hold the tester by its insulated handles.*

- If the meter indicates an incorrect voltage for any test, turn off the power and replace the outlet, connecting wires to the new outlet as you found them on the old one.
- Restore power and test the new outlet for voltage.

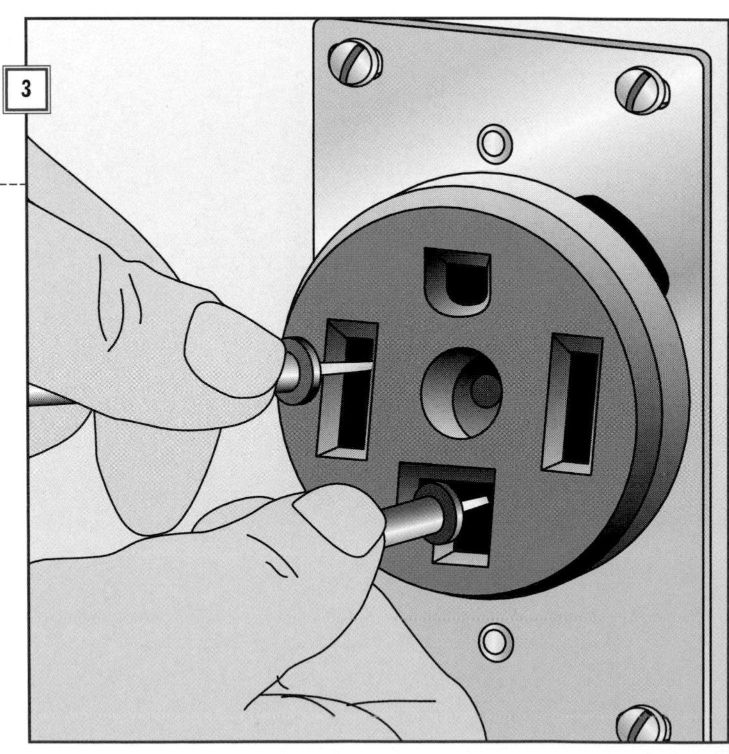

# IDENTIFYING THE WIRING IN A BOX

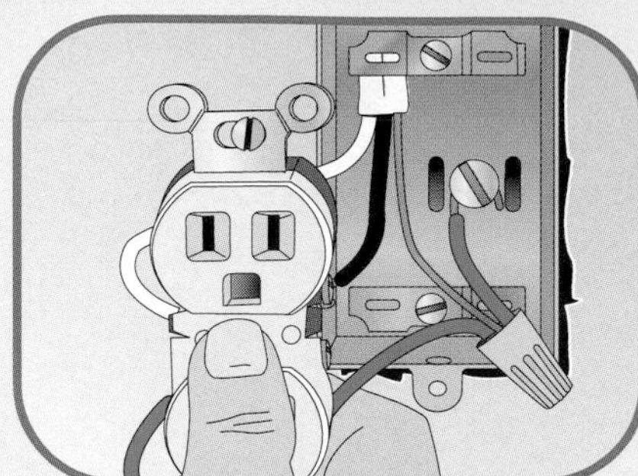

END-OF-RUN OUTLET

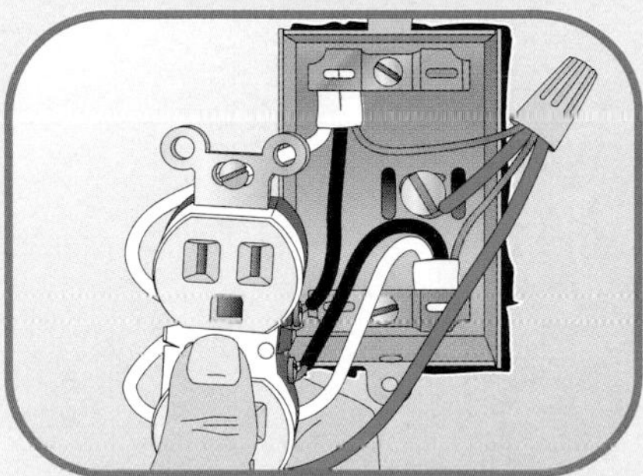

MIDDLE-OF-RUN OUTLET

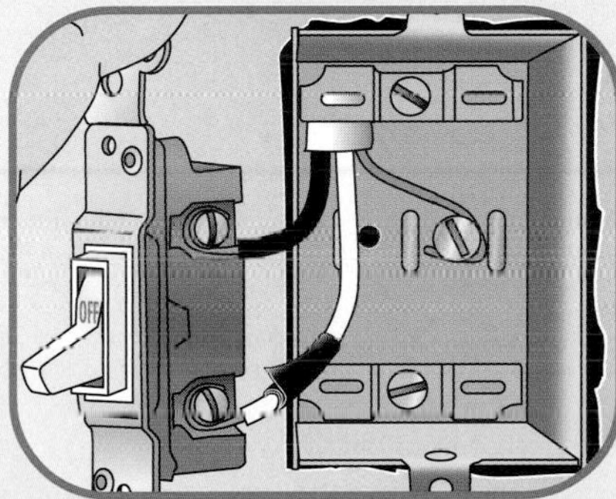


END-OF-RUN SWITCH

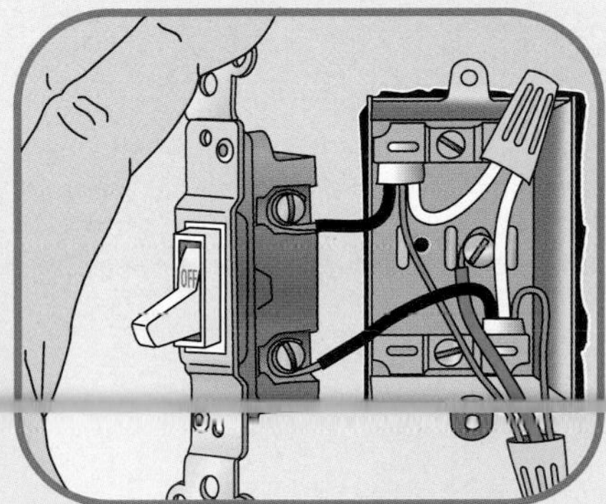


MIDDLE-OF-RUN SWITCH

When you open an outlet or switch box, it can be useful to figure out the position of the outlet or switch in the circuit, as well as the function of each wire. This knowledge can help you pinpoint problems and connect wires to the correct terminals when making repairs.

If you can't find the source of a problem with an outlet or a switch, work from that point back to the service panel, troubleshooting each load on the circuit and its connections until you locate the fault.

**End-of-run outlet:** When there's only one cable entering an outlet box, it means the outlet is the last fixture on the circuit. Power comes from the service panel along the black (hot) wire through other outlets, switches, and light fixtures on the circuit and begins its return to the source through the white (neutral) wire attached to this outlet. The black wire attaches to a brass terminal, the white wire, to a silver terminal. The grounding pigtail has three wires, with jumpers connected to the outlet's grounding terminal and to the box.

**Middle-of-run outlet:** Two cables entering an outlet box indicate that the outlet is not the last fixture on a circuit. One of the black wires receives power from the service panel; the other sends it on to other loads on the circuit. The white wires allow current passing through the outlet and other loads on the circuit to return to the panel. Make sure a jumper from the outlet's ground terminal is connected to the grounding pigtail.

**End-of-run switch:** Switches have only hot wires connected to them; hot wires are usually black, but they may also be red, or white marked with black tape or paint. When there's only one cable in the switch box, it's at the end of the run. One of the wires connected to the switch receives current from the source, and the other passes it on to the light or other load when the switch is flipped on. Current returns to the source through through the load's white wires. Always make sure that a switch is grounded in one of the ways shown on page 81.

**Middle-of-run switch:** When there are two cables in a switch box, it's in the middle of the run. The black wires are connected to the switch terminals, and the white wires are joined with a wire cap to complete the circuit. A middle-of-the-run switch must also be grounded by one of the methods shown on page 81.

# FIX IT: Extending Circuits

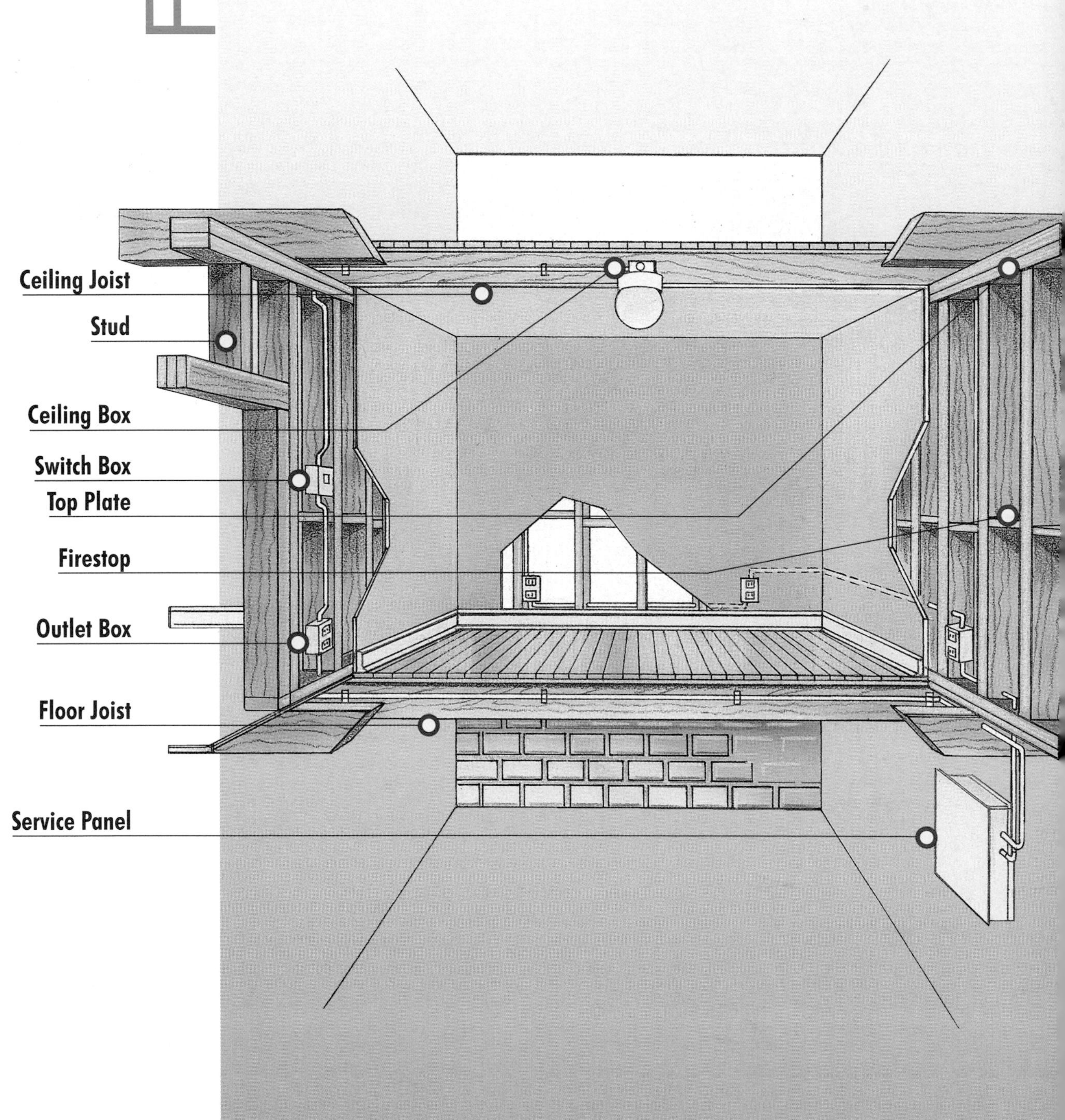

# Chapter 6

## Contents



## How It Works

**The illustration at left shows several ways to run electrical cable through a home. The simplest route begins at an existing outlet and leads either through an unfinished basement (if the outlet is on the ground floor) or through an unfinished attic (if the outlet is upstairs). Or you can try concealing cable for a circuit extension behind baseboard molding *(page 111)*. These paths generally result in the fewest holes in walls and ceilings that would need patching later.**

**When you extend a circuit between finished floors, holes and patches are generally unavoidable. You'll want to repaint the entire room after you have finished the electrical work.**

# Troubleshooting

| Problem | Solution |
| --- | --- |
| • Not enough outlets in a room; outlet locations inconvenient | Extend circuit through unfinished basement ceiling **104** •<br>Extend circuit through unfinished attic **108** •<br>Extend circuit behind baseboard molding **111** •<br>Extend circuit with surface-mount wiring system **117** • |
| • Not enough overhead lights in a room; ceiling box location inconvenient | Run cable for a new ceiling box **113** • |
| • Not enough switches in a room; switch location inconvenient | Install a new wall box **113** • |
| • Framing blocks the route for new cable | Cut inspection hole in wall **116** • Notch passage in firestop **109** •<br>Notch passage in top plate **114** • |
| • Masonry construction prohibits running cable within walls | Extend circuit with surface-mount wiring system **117** • |
| • Confusing wiring connections at box where circuit extension begins | Identify the incoming hot wire **120** • Determine whether circuit extension occurs at the end or middle of the run **120–121** • |
| • Hazardous wiring condition caused by mismatched cables | Replace inadequate wiring with wiring of correct type and gauge **105** • |

# Before You Start

Extending circuits isn't difficult, but it can be time-consuming. And you'll need a helper when fishing cable through walls and floors.

### ADD UP AMPS BEFORE ADDING ON

Before running new cable and adding loads to a circuit, calculate the existing load and confirm that the wiring and circuit breakers or fuses have sufficient capacity *(page 13)*. If capacity is insufficient, extend another circuit that has excess capacity, or have an electrician add another breaker or fuse to the panel or otherwise upgrade electrical service.

When you're running new cable inside a wall or floor frame, conditions sometimes prevent you from recessing it the required 1-1/4 inches from the framing face. In such cases, electrical codes require that the cable be protected by a steel safety plate to prevent nails and screws from accidentally puncturing the conductors. Safety plates, available at hardware stores, are at least 1/16 inch thick and are made with sharp points so you can easily fasten them to studs with a hammer *(page 115)*.

## Before You Start Tips:

- Outlet boxes tend to be better sources for extensions than switch boxes—the connections are usually simpler.
- If you extend a circuit from a switch box, the switch will control the extension if the box has only one cable leading to it.
- Make sure the box at the beginning of an extension is large enough for the number of wires it will ultimately contain *(pages 120–121)*. If not, replace it with a larger box.

Hammer
Tape measure
Fish tapes
Electric drill
1/4-inch drill bit
3/8-inch drill bit
3/4-inch spade bit
Drill-bit extension
Utility knife
Wood chisel
Pliers
Long-nose pliers
Cutting pliers
Cable stripper
Multipurpose tool
Flat pry bar
Hacksaw
Screwdrivers
Flashlight

Cable
Cable staples
Electrical boxes
Electrical tape
Masking tape
Wire caps
Switches
Outlets
Steel safety plates

Before opening an electrical box, turn off power to the circuit at the main panel and perform a voltage test *(page 20)*.

Extend a circuit only with cable that matches the gauge of the existing wiring *(page 105)*. Using smaller cable creates a fire hazard.

# Running Cable across a Basement

## 1. Preparing for the New Run

To extend a circuit in a room above an unfinished basement, run cable down from an existing box, along a joist or across the joists, then up to the new location *(right).*

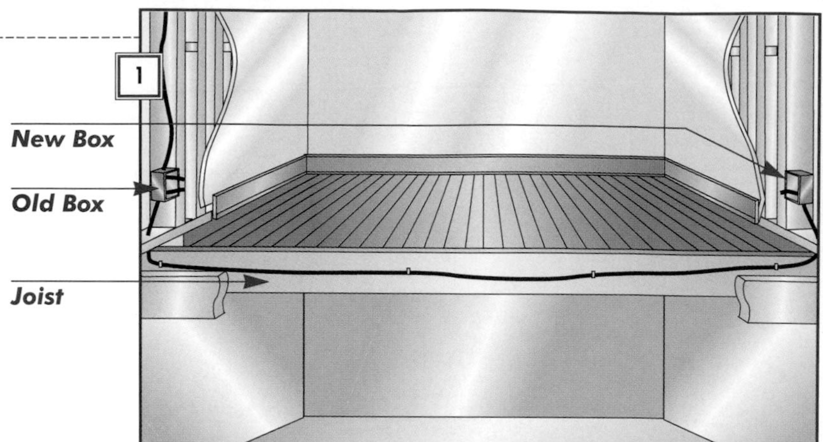


- Turn off power to the circuit at the service panel *(page 10).* Test for voltage at the box where you plan to begin the extension, to confirm that the circuit is dead *(page 20).*

- Disconnect and remove the switch *(page 70)* or outlet *(page 90)* from the box.

- Confirm that the box is big enough to hold extra wires *(page 121);* if not, replace it.

- Cut a hole for the new box *(page 116).*

- To find the length of cable you need, measure from the old box to the new one, then add 20 percent.

## 2. Boring a Location Hole

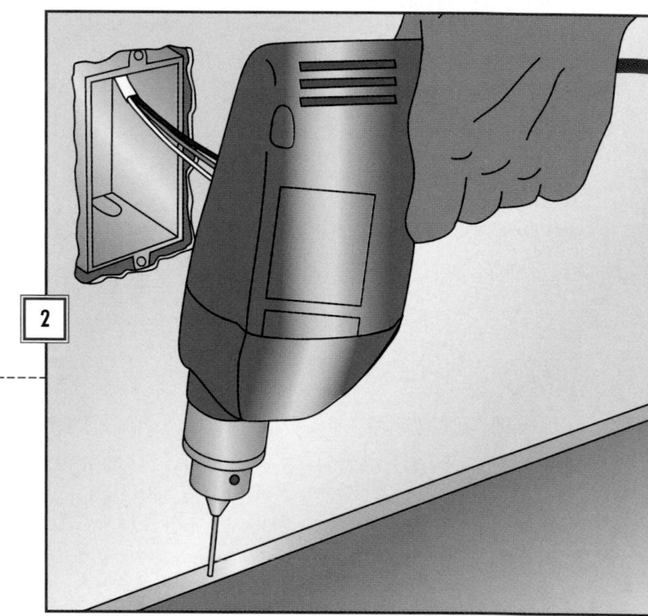


- Remove the section of baseboard below the existing box and below the hole for the new one *(page 111).*

- Drill 1/4-inch-diameter location holes through the floor, one directly below the box *(right),* the other below the new hole.

- Insert a stiff wire in each hole; bend it slightly to prevent it from slipping through.

## 3. Creating a Passage for the Cable

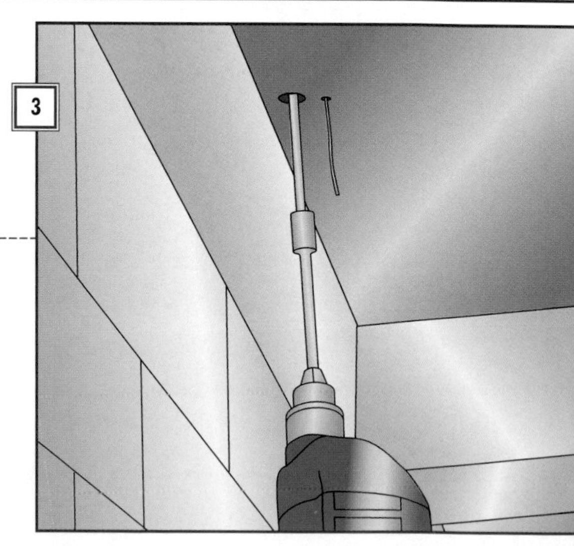


- Find the location wires in the basement.

- From each wire, measure 2 inches toward the wall and make a mark.

- Fit a drill with a drill-bit extension and a 3/4-inch spade bit. At the marks, bore upward through the subfloor and the wall sole plate *(right).*

## CABLE BASICS

A cable consists of two or more wires wrapped in a protective sheath. Insulated wires inside a cable are called conductors to distinguish them from the bare ground wire, which normally carries no electricity.

Romex cable is the type most commonly found in houses. It comes in two kinds of sheathing: NM, which means that the sheathing is flame-retardant and moisture-resistant; and NMC, which is fungus- and corrosion-resistant. NMC is commonly used in damp areas, like basements.

Other markings on the cable sheathing tell what's inside. In the top example at right, 14-2G means that the cable contains two 14-gauge insulated conductors.plus a ground wire. As shown in the chart below, this cable can safely carry 15 amps of current.

The cable in the center photo is 12-gauge, three-conductor Romex with a ground wire. Although the gauge number is smaller than in the first example, the conductors are thicker and can safely handle up to 20 amps. Often used in its two-conductor version in 120-volt circuits serving the kitchen and other parts of the house where demand for electricity is high, three-conductor cable of this capacity appears in 120/240-volt circuits.

Flexible armored cable, or BX *(bottom)*, has a tough metal sheathing instead of vinyl. Shown here in a 12-gauge, two-conductor version, BX is used in walls framed with metal studs and in exposed locations where there's a risk of damage—a garage, basement, or workshop, for example. It is suited only to dry areas. In BX cable, a bare bonding wire acts as a ground wire; fasten it to the grounding screw in the box.

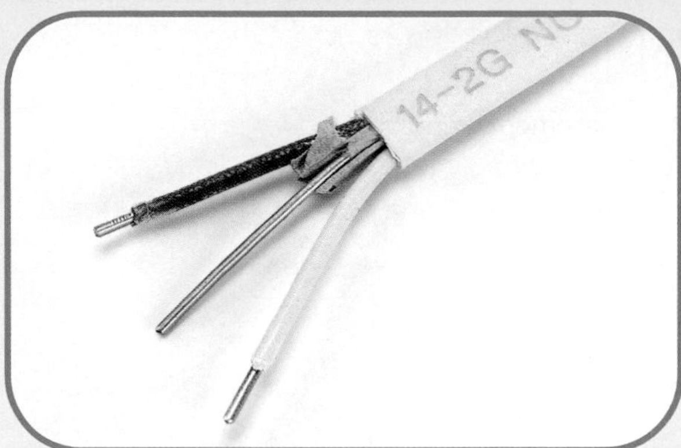


***NO. 14, 2-CONDUCTOR NM CABLE WITH GROUND***

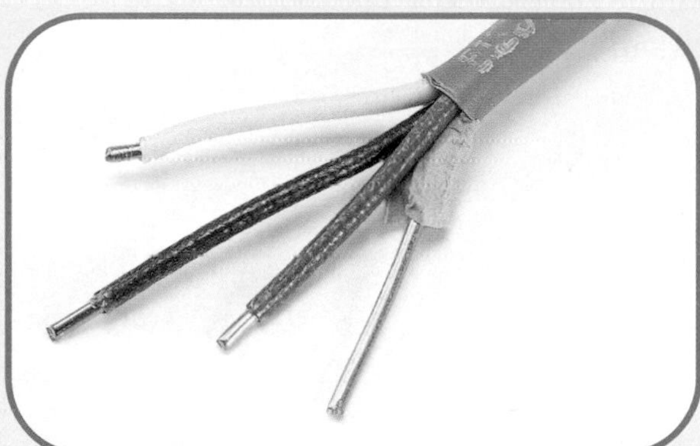

***NO. 12, 3-CONDUCTOR NM CABLE WITH GROUND***

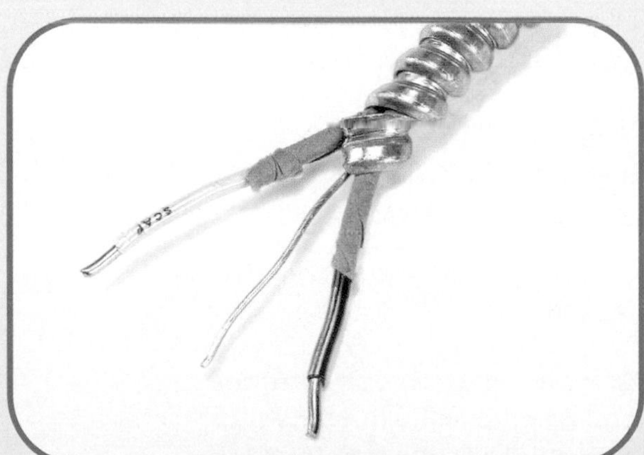

***NO. 12, 2-CONDUCTOR BX CABLE***

### WIRE GAUGE AND AMPERAGE

| **WIRE GAUGE** | No.6 | No.8 | No.10 | No.12 | No.14 | No.16 | No.18 |
|---|---|---|---|---|---|---|---|
| **AMPERES** | 55 | 40 | 30 | 20 | 15 | 10 | 7 |

## 4. Feeding fish tape to the basement

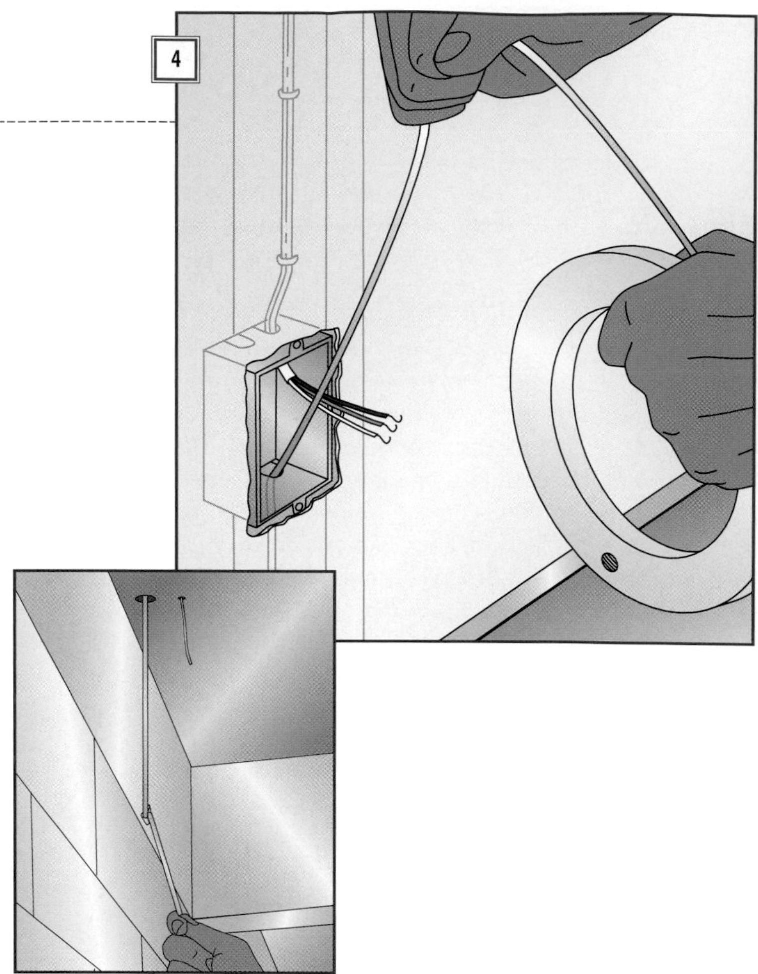


- Temporarily remove the internal cable clamp from the bottom of the box where your run begins, and then open a knock-out hole.
- Have a helper in the basement guide a fish tape into the hole in the subflooring
- Open the hook of another fish tape and feed it from the box toward the basement *(right).* Move the tape back and forth in the wall cavity to snag the other fish tape.
- Have your helper pull the tapes down into the basement *(inset).*

## 5. Pulling up the cable

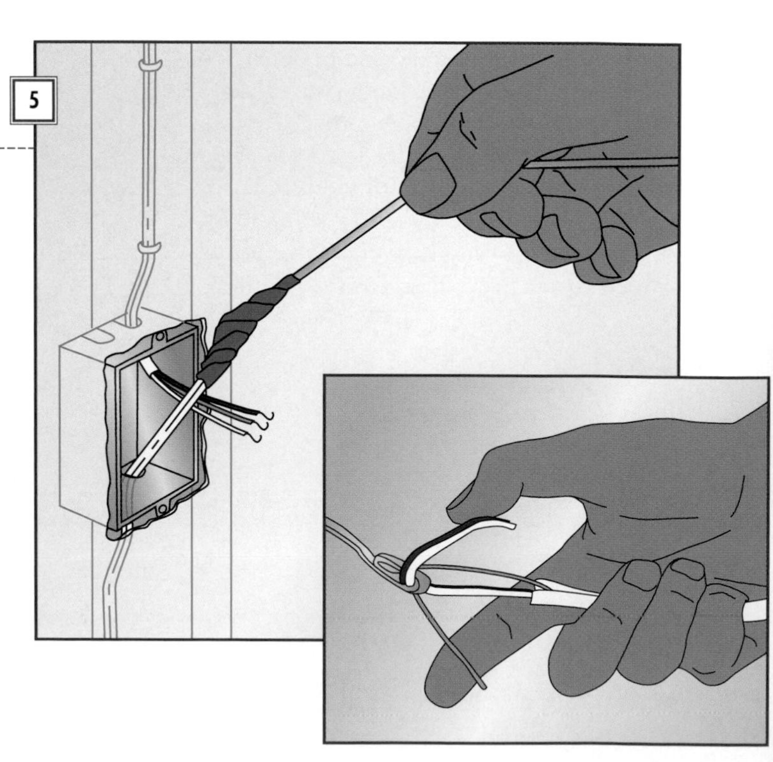


- In the basement, strip back 8 inches of sheathing from the new cable *(page 110).*
- Thread the wires through the hook *(inset)* and wrap with tape.
- Have your helper feed cable from the basement as you pull the fish tape up through the wall into the box *(right).*
- Detach the cable from the fish tape, and secure it to the box by reinstalling the internal clamp.

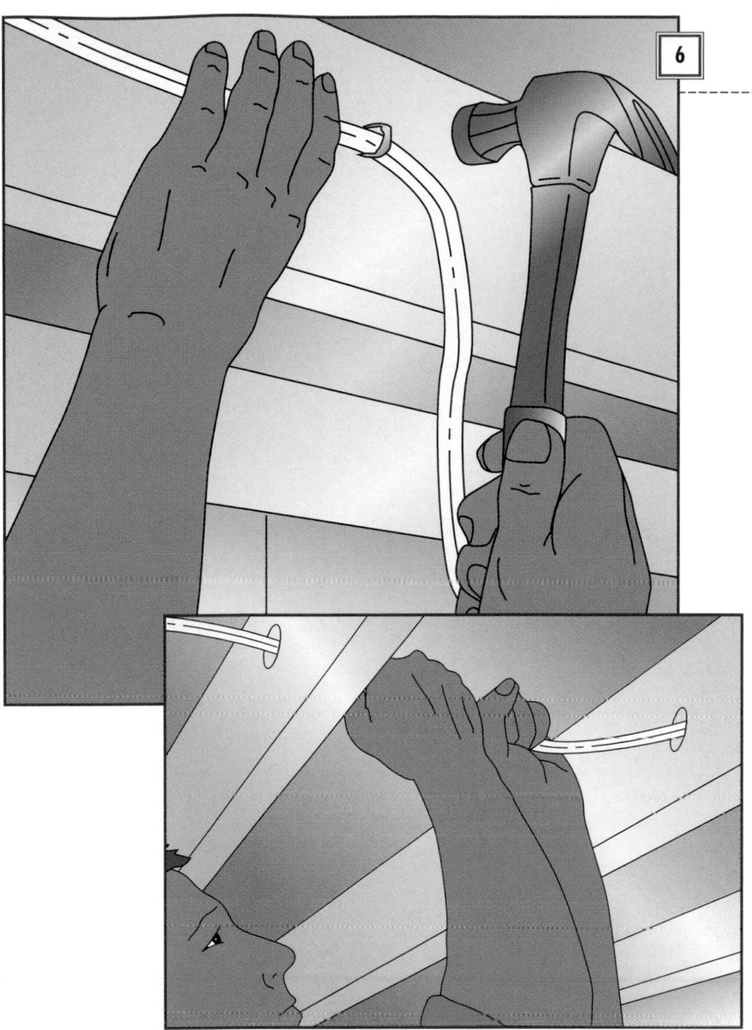

## 6. Running cable along or through joists

- If possible, run the cable along the side of a joist, securing it every 4-1/2 feet with cable staples *(left).* Be careful not to damage the cable as you are hammering the staples: It takes very little pressure to cut into the plastic sheathing.

- To run cable across joists *(inset),* first bore a series of holes where needed with a 3/4-inch spade bit. Drill as close to the center of the boards as possible—a hole too close to the top or bottom of a joist weakens it. Pass the cable through each hole.

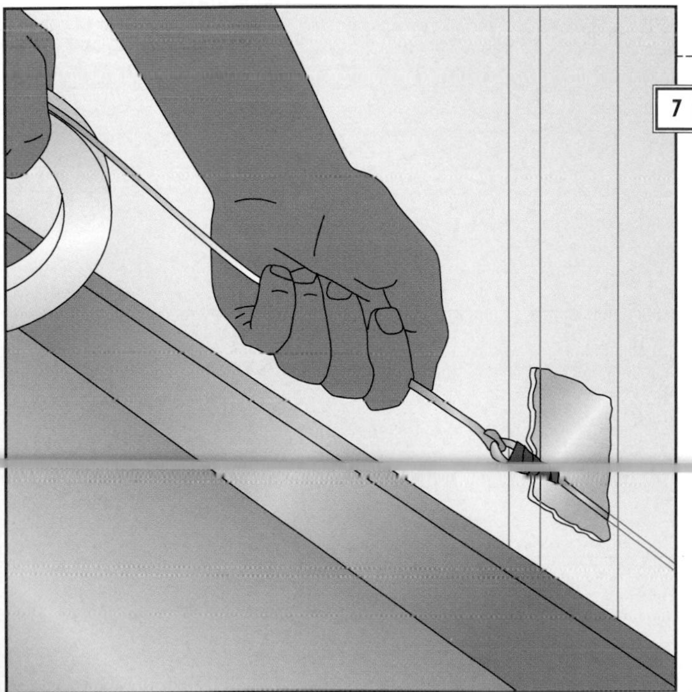

## 7. Pulling cable to the new location

- Fish the cable through the hole for the new box *(left).*

- Thread the cable though a knockout hole in the new box. Clamp the cable securely and trim it, leaving 8 inches for connections. Then install the new box in the wall *(page 116).*

- Connect the new switch *(page 70)* or outlet *(page 90),* and install it in the box.

- Return to the existing box and make the necessary connections to restore power to the old outlet or switch and provide power to the new device *(pages 120–121).*

- Turn on the power.

# Routing Cable through an Attic

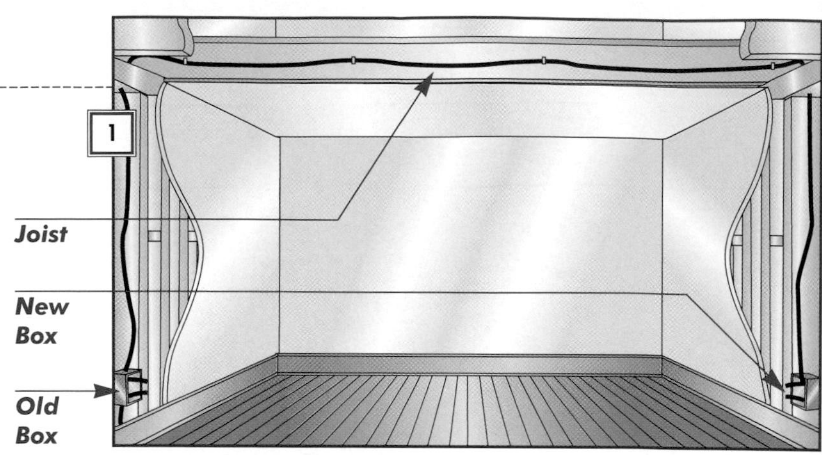


## 1. Preparing for the Run

If there's an unfinished attic above a room in which you wish to extend a circuit, the best approach is usually to run cable up from an existing box, along the ceiling joists, then down to the new location *(right).* This method causes minimal damage to wall and ceiling finishes.

- Cut power to the circuit at the service panel, and begin the job as described on page 104, Step 1.

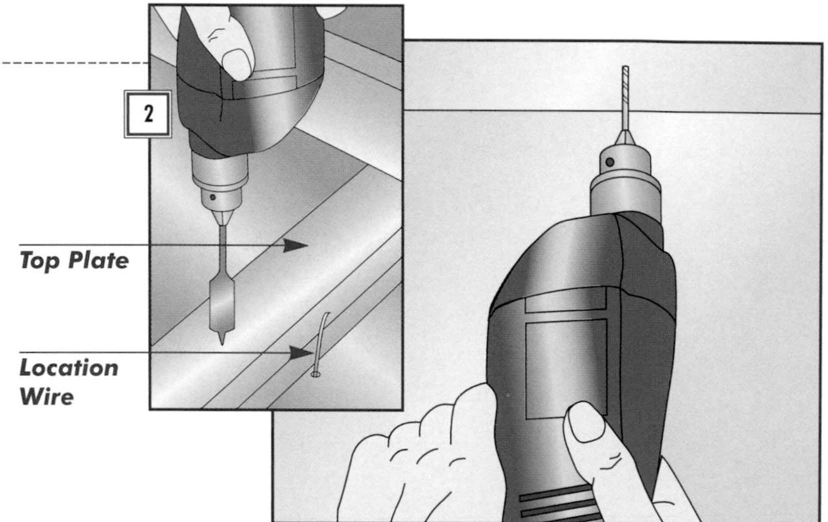


## 2. Drilling Location and Cable Holes

- Drill 1/4-inch-diameter location holes through the ceiling directly above the existing box and above the spot for the new box.
- Bend two 12-inch lengths of stiff wire to a gentle curve, then push them up through the location holes.
- In the attic, lay planks across the exposed joists so you can move around more easily.
- Find the location wires. Using them as guides, bore 3/4-inch cable holes down through the center of the wall top plate *(inset)* and into the wall cavity. If you can't drill through a plate from above, notch it from below *(page 114).*

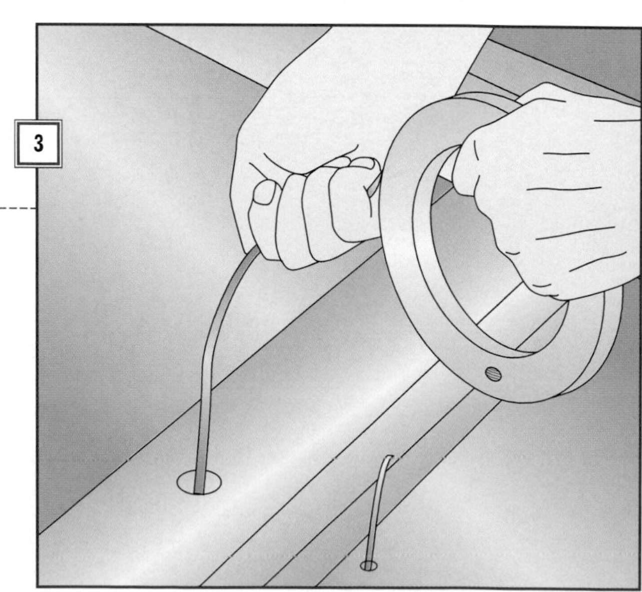


## 3. Feeding the Fish Tape Down from the Attic

- Working from the attic, feed the fish tape down through the hole above the existing box *(right).*
- If the tape hits an obstruction, shake or snap it free. Whenever that technique fails, the blockage is probably a firestop, which you will have to notch *(Step 4).*

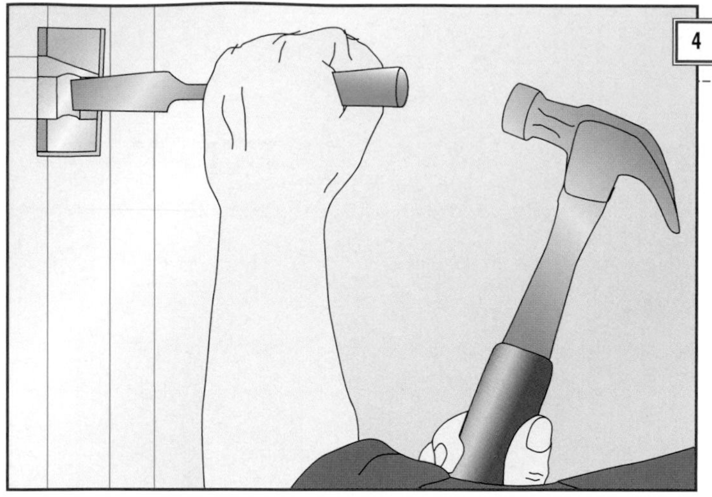

## 4. Notching a Firestop

- If the tape seems obstructed by a blocking member, mark the tape, pull it out of the wall, and measure along the tape to find the distance to the obstruction.

- Remove a piece of drywall at the spot *(page 116),* then chisel a notch *(left).* After fishing cable through the notch *(Step 5),* fasten a 1/16-inch steel plate over the notch and patch the hole *(page 115).*

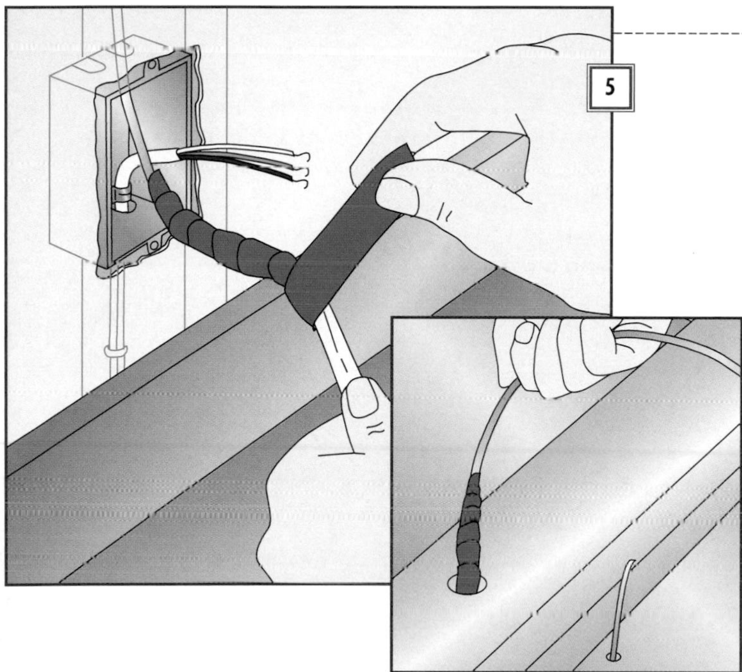

## 5. Fishing the Cable Up Through the Existing Box

- Have a helper feed a second fish tape up through the box while you attempt to hook it from the attic with the first tape. Then ask your helper to pull the ends of both tapes into the room and disconnect them.

- Strip 8 inches of sheathing from the new cable *(page 110).*

- Thread the wires through the hook, and tape the connection *(left).*

- Pull the cable up to the attic with the fish tape *(inset)* while your helper feeds it from below. Leave about a foot of cable protruding from the box.

- Fasten the cable to the existing box with the internal clamp.

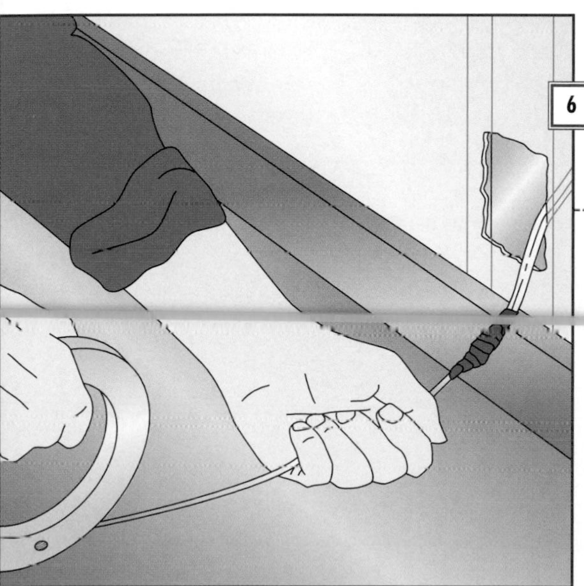

## 6. Fishing Cable to the New Location

- Fish the cable down to the new location the way you fished it up to the attic.

- Pull the cable through the hole you cut in the wall *(left).* Clamp the cable in a new box and install the box in the wall *(page 116),* then connect the new switch *(page 70)* or outlet *(page 90).*

## WIRE-CUTTING AND STRIPPING TECHNIQUES

Preparing cable for use entails three operations: cutting the cable, removing the sheathing, and stripping insulation from the individual conductors. There's more than one way to cut and strip each type of wire and cable. The tools and techniques shown here represent some of the most efficient methods.

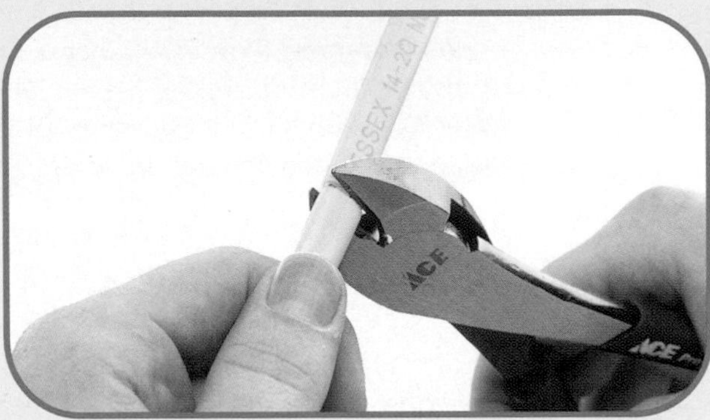


**Cutting Romex cable.** Place the cable in the jaws of a pair of diagonal-cutting pliers, and squeeze sharply. Always cut cable longer than needed. It's easy to shorten a cable, but impossible to lengthen one.

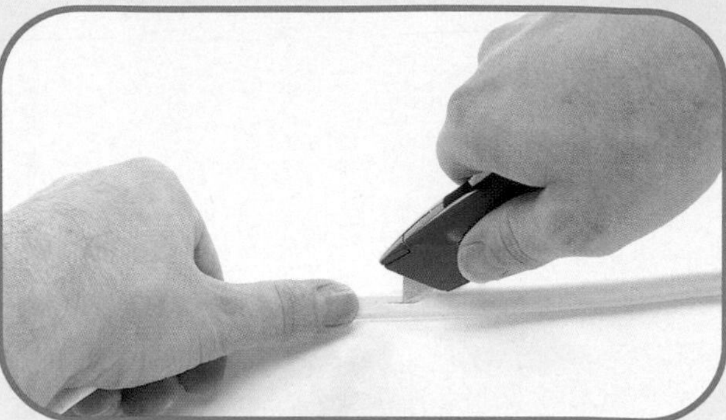

**Removing sheathing with a utility knife.** Center the blade of a utility knife on the sheathing 8 inches from the end of the cable. Draw the knife along the cable, taking care not to nick the wires inside. Peel back the sheathing and cut it off.

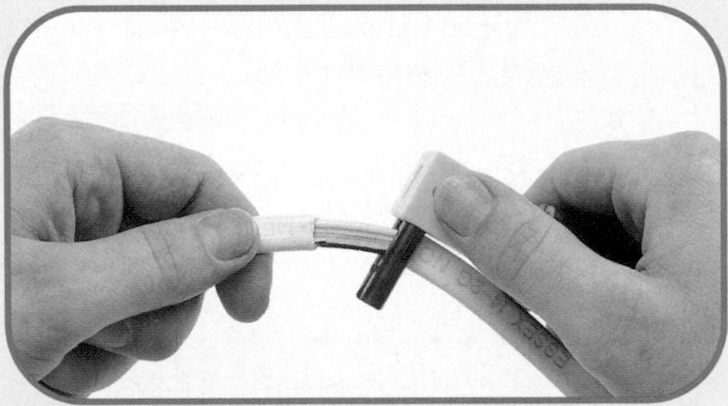

**Using a cable stripper.** There are several kinds of cable strippers. To operate the kind shown here, open the jaws, insert the cable about 8 inches into the jaws, then close them. Turn the tool around the cable to cut the sheathing, then slide it off the cable.

**Cutting BX cable.** Bend BX cable to spread the spirally wound sheathing. Cut the sheathing with diagonal-cutting pliers, then sever the wires inside the cable, including the bonding wire. To remove sheathing, repeat the process without cutting the wires.

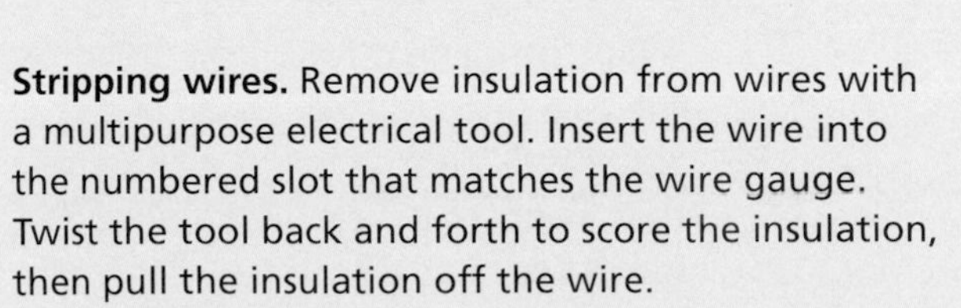

**Stripping wires.** Remove insulation from wires with a multipurpose electrical tool. Insert the wire into the numbered slot that matches the wire gauge. Twist the tool back and forth to score the insulation, then pull the insulation off the wire.

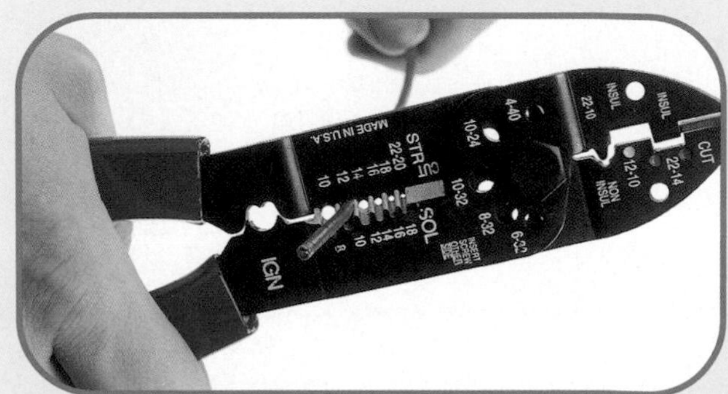

# Hiding Cable behind a Baseboard

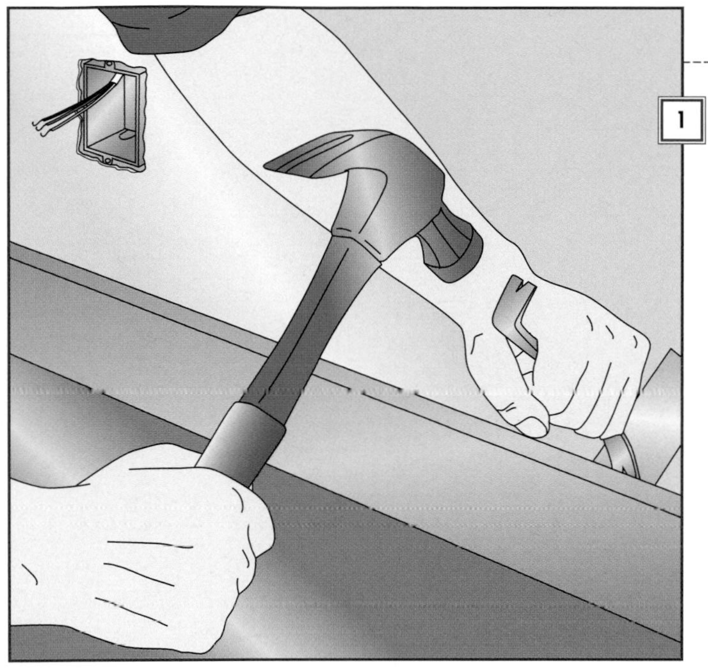

1

### 1. Prying off the baseboard

This technique works best if baseboard molding is more than 2-1/2 inches tall.

• Turn off power to the circuit at the service panel *(page 10)*. With a multitester, perform a voltage test *(page 20)* at the box from which you plan to run the new cable, to confirm that the circuit is dead.

• Cut a hole for the new box *(page 116)*, and calculate the length of wire you need *(page 104)*.

• Use a hammer and flat pry bar to remove any shoe molding and the baseboard between the existing box and the opening for the new one *(left)*. A piece of scrap wood behind the bar will keep it from damaging the wall.

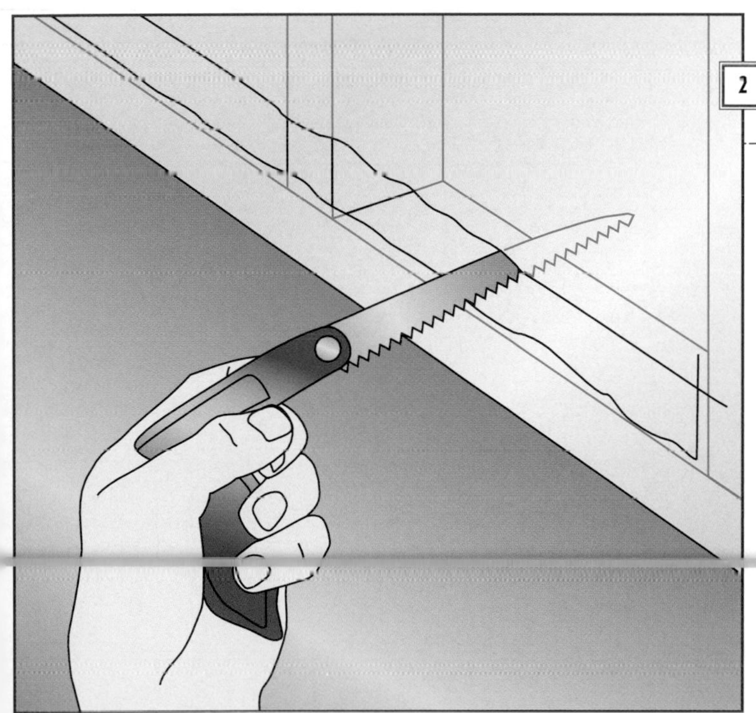

2

### 2. Cutting a channel

• At a point at least 1-1/2 inches above the floor and below the top of the baseboard, cut a channel about 1 to 1-1/2 inches wide, using a keyhole saw *(left)*. Punch a hole in the wallboard to admit the tip of the saw blade; at studs, cut the wallboard with a utility knife.

### 3. Routing cable through studs

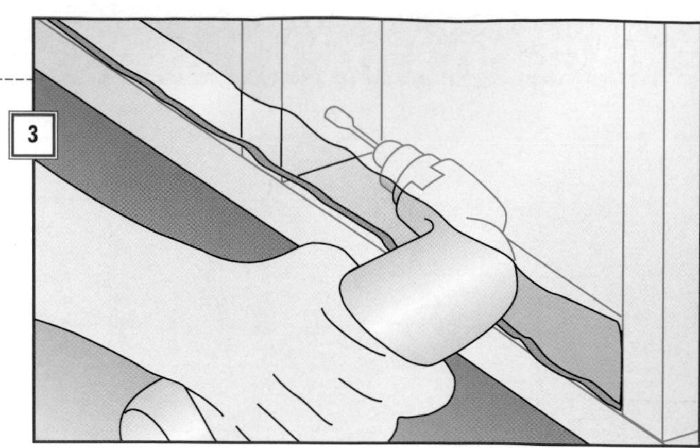


- Drill 1/4-inch-diameter location holes.
- Fit a right-angle drill, available from tool-rental stores, with a 3/4-inch spade bit. Bore a hole in each stud along the cable run at least 1-1/4 inches back from the front edge *(right).*
- Alternatively, chisel notches 1/2 inch deep in the edge of each stud along the cable run.

### 4. Fishing cable to the existing box

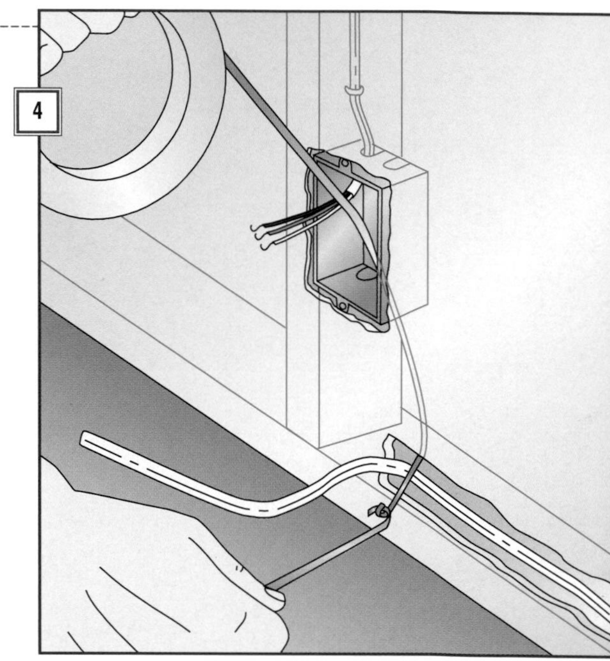


- Remove the internal cable clamp from the bottom of the box, then feed a fish tape through a knockout hole until it strikes the wall's sole plate.
- Insert the second tape with an opened hook into the channel.
- Snag one tape with the other, then pull the first into the channel with the second *(right),* and disconnect the tapes.
- Strip the sheathing, then fasten the wires to the fish tape *(page 106)* and pull the cable into the box.
- Detach the wires from the fish tape and clamp the cable to the box.

### 5. Fishing cable to the new location

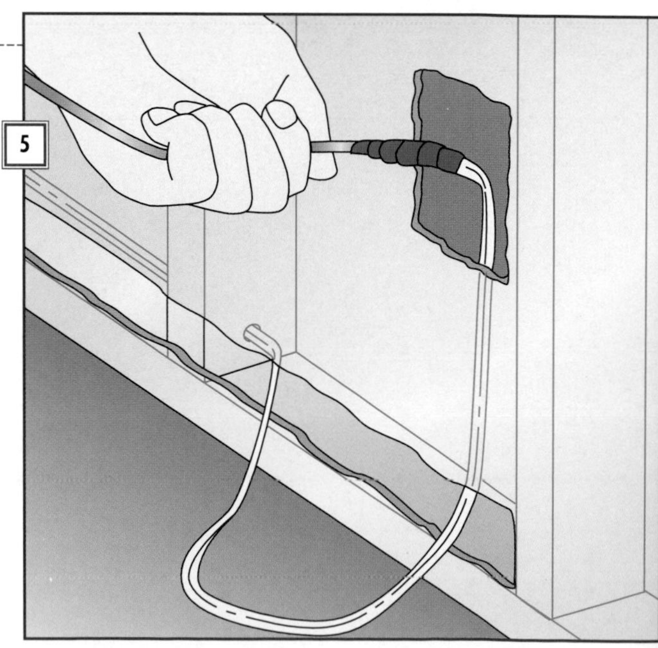


- Thread the cable through the studs, and fish it through the opening you made for the new box *(right).* If you cut notches instead of drilling holes, hammer a steel safety plate *(page 103)* over each notch. Between studs, bend the cable and staple the cable at least 1-1/4 inches back from the sole plate's edge.
- Thread the end of the cable through a knockout in the new box, then install the box *(page 116).* Clamp the cable in the box, and install the new switch *(page 70)* or outlet *(page 90).*
- Restore power and test the new outlet.

# Wiring for an Overhead Fixture

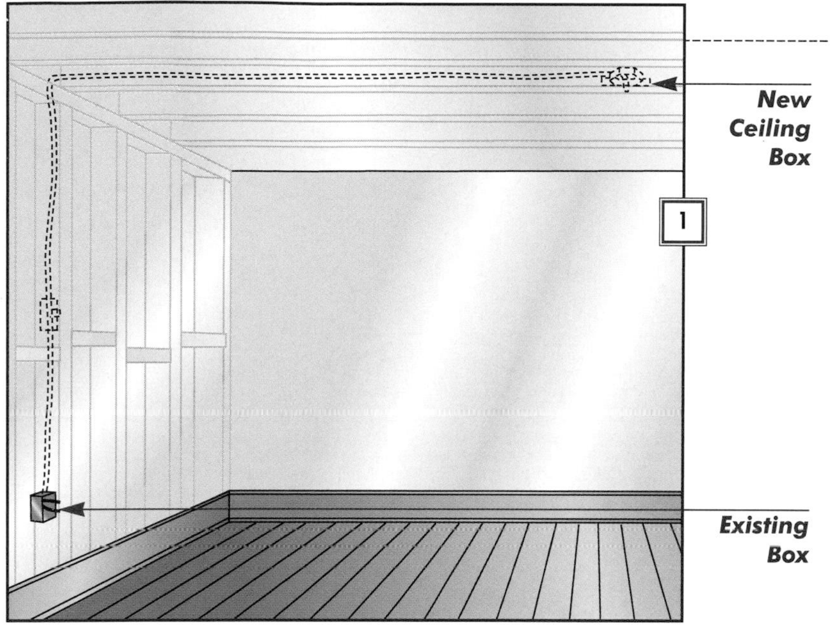


## 1. Preparing for the run

A circuit extension below a finished space *(left)* goes from an existing outlet box to a new ceiling box. For simplicity, the switch is shown above the old box instead of near a door, a location that requires routing the circuit first to the switch using techniques shown on pages 111and 112. These methods also apply when offsetting the ceiling box from the wall box.

- Mark the location for the new ceiling box, and with a keyhole saw, cut a hole for it there. Shine a flashlight into the hole to determine the direction of the joists.

- Choose an existing box on a wall at the joist ends from which to extend the circuit. Turn off power at the service panel *(page 10)*, and test voltage at the box to confirm that power is off *(page 20)*.

- Disconnect the outlet or switch. Make sure that the box is large enough to add more cable; if not, replace it with one of the appropriate size *(page 121)*.

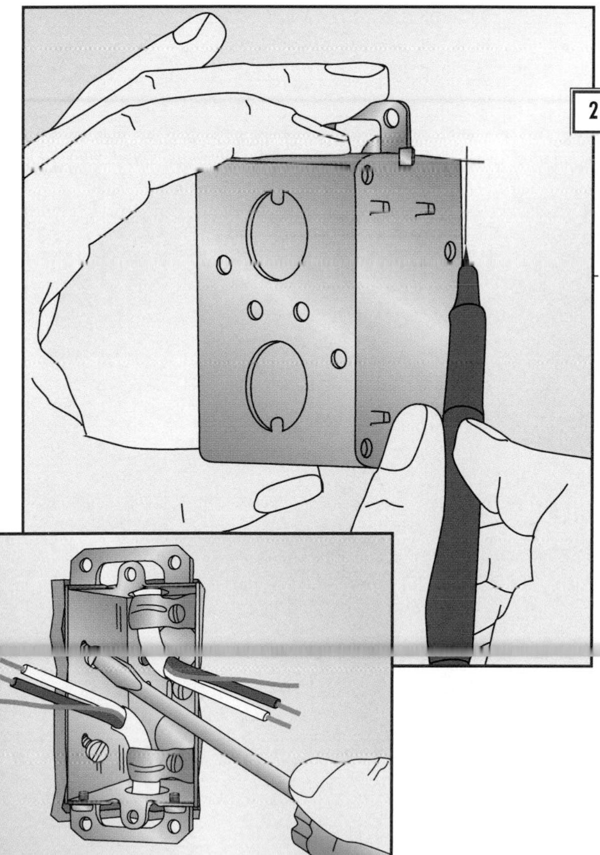


## 2. Installing a wall box

- Locate a stud closest to the place where you want to install the new switch. On the wall finish, mark the location at the edge of the stud.

- Set the box on the mark against the wall finish and trace its outline *(left)*.

- Drill a 3/8-inch hole at each corner of the outline, then cut out with a keyhole saw.

- Fish the cable to the opening from the the existing box *(page 109)*. Insert it through a knockout in the new box, and secure it with the box's internal clamp. Set the switch box in the hole so it's flush with the wall. Fasten it to the stud with screws *(inset)*.

### 3. Notching the Top Plate

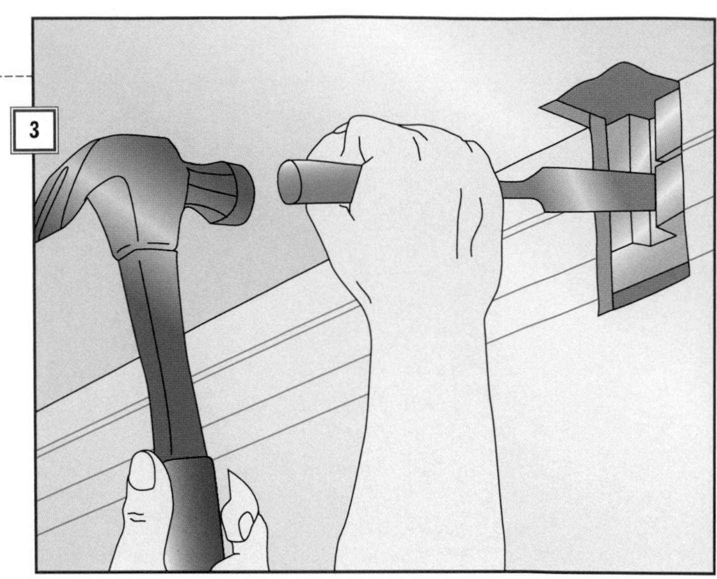


Where the cable will cross from wall to ceiling, cut an L-shaped opening in the wall and ceiling to expose the wall top plate. Use a utility knife to cut drywall, or a cold chisel and wood chisel to cut through plaster and wood lath.

- With a wood chisel, cut a notch about 3/4 inch wide and 1/2 deep in the wall top plate to form a passage for the cable *(right).*

### 4. Fishing Cable from the Switch Box

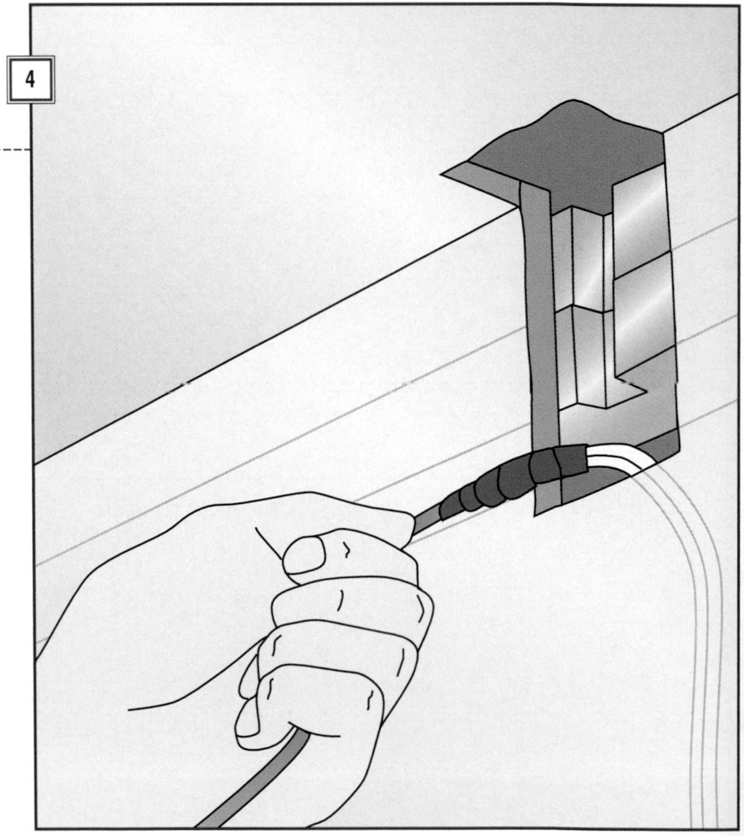


- Measure the cable route from the switch box to the hole for the ceiling box. Add 20 percent and cut a piece of cable to this length *(page 110).*

- Using the techniques from page 109, fish the new cable from a knockout hole at the top of the switch box to the top-plate notch *(right).*

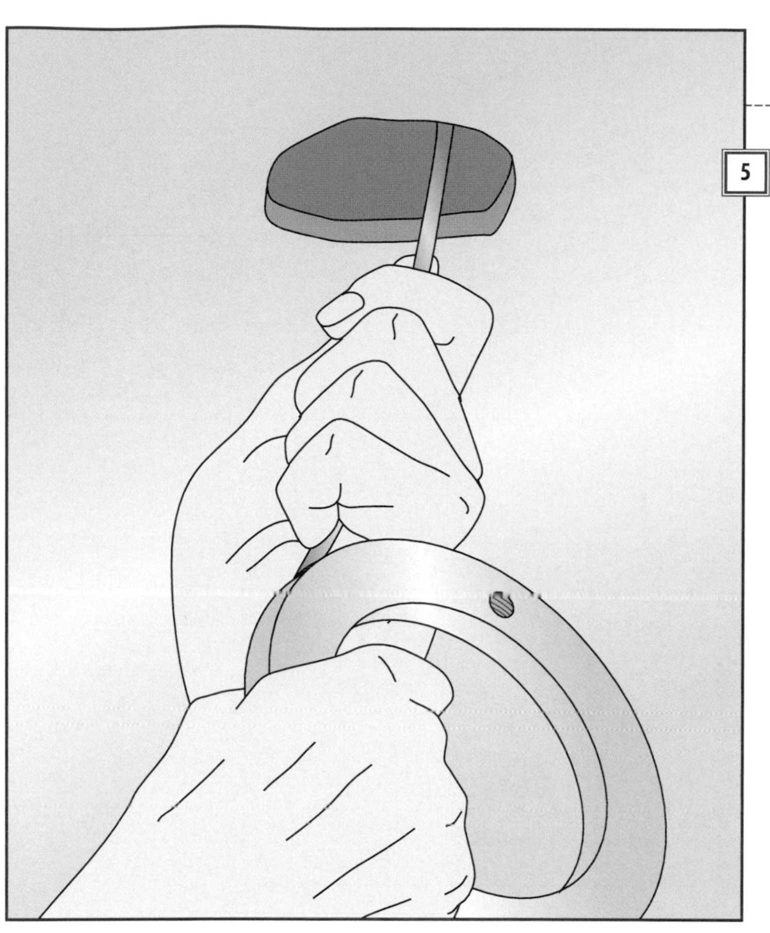


### 5. Fishing cable across the ceiling

- Guide a fish tape from the ceiling-fixture hole toward the top-plate notch *(left).*
- Have a helper open the hook of a second fish tape and insert it into the notch to snag the first fish tape; pull both tapes back through the notch.
- Detach the second tape and fasten the cable to the tape running through the ceiling.
- Pull the fish tape through the hole at the new ceiling-box location, bringing the cable with it.
- Cut the cable with diagonal-cutting pliers, allowing 8 inches of wire to make connections to the light fixture.

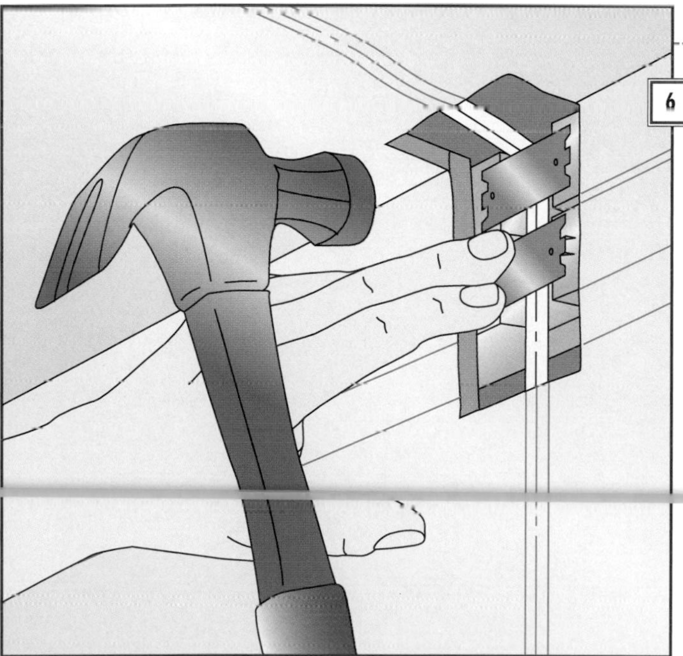


### 6. Protecting the cable

- Hammer a steel safety plate into each board of the wall's top plate to cover the notch where the cable passes through the plate and to protect the cable *(left).*

### 7. Installing the Hanger Bar and Ceiling Box

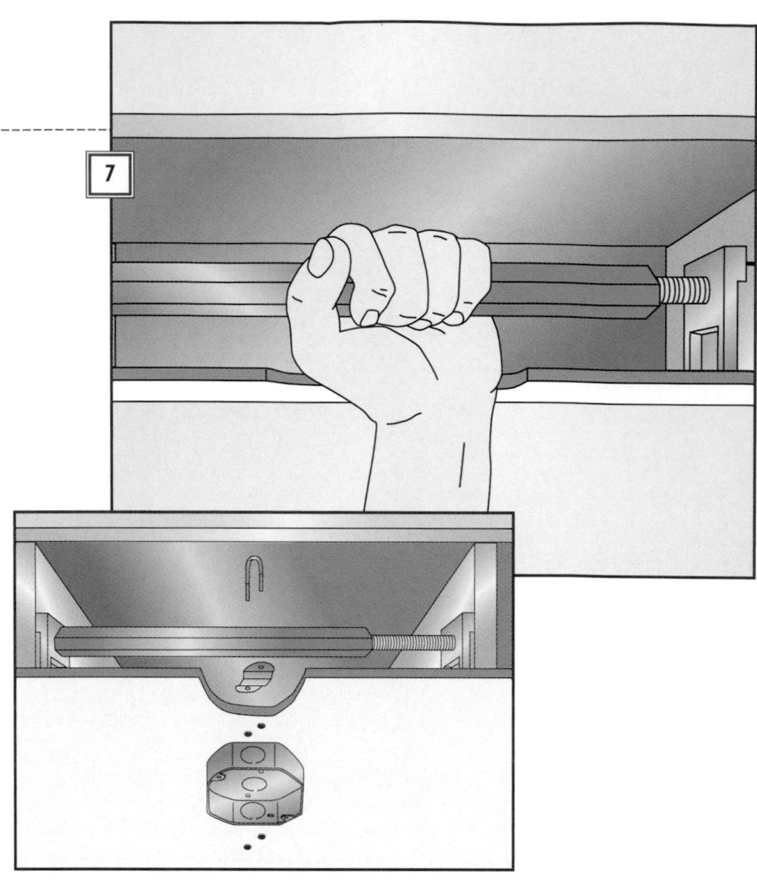


- Place the ceiling box over the hole you cut in the ceiling in Step 1 *(page 113)*, and mark an outline of its exact shape on the ceiling. Then with a keyhole saw or utility knife, cut along the marks.
- Slip a screw-type hanger bar (available at home centers and electrical supply houses) into the hole and rest its feet on the top surface of the ceiling. Making sure the bar is perpendicular to the joists, turn the outer sleeve clockwise to extend the bar until the ends touch the joists *(right)*.
- With an adjustable wrench, continue turning the sleeve to force the bar's teeth firmly into the joists.
- Hang the U-bolt over the bar and fasten the box with the strap, washers, and nuts provided, as shown *(inset)*.
- Make the necessary connections at the outlet box and switch box *(pages 120–121)*. Make the connections at the ceiling box, and mount the fixture *(see Chapter 3)*.

## Ron's TRADE SECRETS

**Cutting down on patchwork**

When I need to cut a small opening in wallboard to fish some cable or peek at wiring, I use this "easy-patch" method. I start by cutting a small rectangle or circle with a keyhole saw, keeping the handle tipped away from the cut line. The result is a panel with beveled edges *(right)* that fits the opening without falling in. When I'm done with the opening, I butter the edges of the cutout with wallboard compound and press it in place. To finish, I smooth a skim-coat of compound over the area.

# Surface Wiring

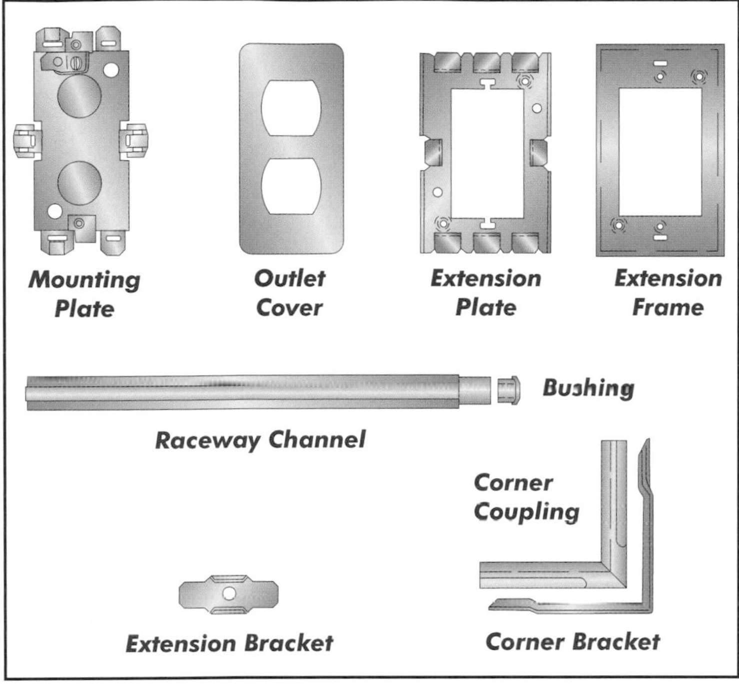


## Surface-mount components

Surface-mount wiring is ideal for extending circuits along brick walls, and in other circumstances that make it impossible to fish cable behind the surface. Shown are components of a metal or plastic raceway system needed to extend a circuit from an existing outlet box to a new outlet *(left).* There are also raceway-system components for mounting wall switches and overhead light fixtures.

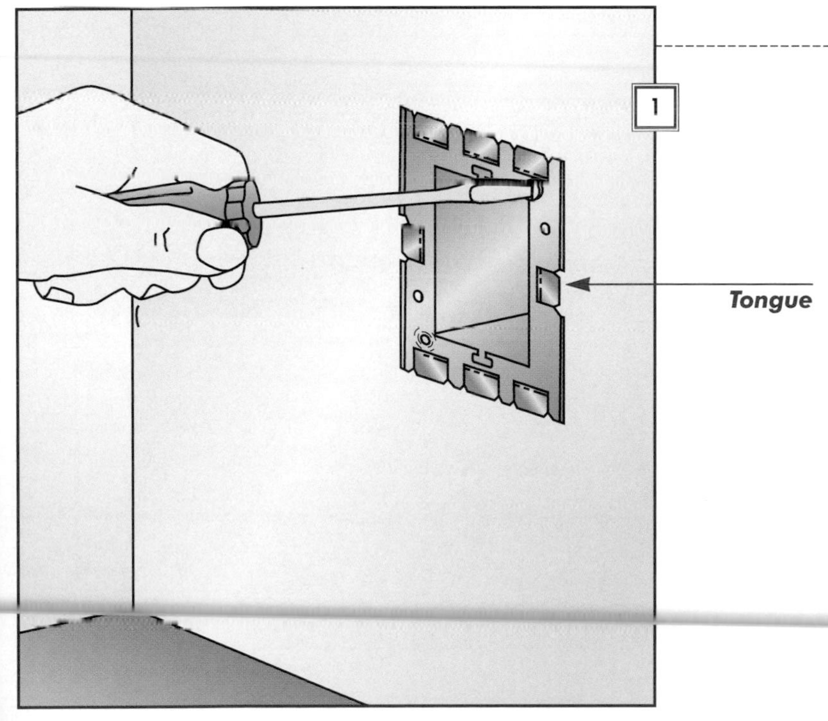


## 1. Mounting the extension plate

- Locate a suitable outlet box from which to begin the circuit extension. Cut power to the circuit at the service panel *(page 10).*
- Remove the outlet cover. Set a multitester to 250 in the AC-voltage range, and test all connections in the outlet box to confirm that the power is off *(page 78).* If there's a voltage reading, return to the service panel and turn off power to the correct circuit.
- Disconnect and remove the existing outlet from the box *(page 90).* Position the extension plate squarely over the box, and drive screws into the slots at the top and bottom to fasten it to the box *(left).*
- Pull the wires out of the box through the extension plate.
- Draw a level line from the tongue on the side of the extension plate to the location for the new outlet.

## 2. Extending the Raceway to a New Location

2

Corner Bracket

Raceway

Mounting Plate

• If the route of the new wiring turns a corner, install the corner bracket with screws and screw anchors, using the level line as a guide. Measure along the line from the extension plate to the new outlet location. Use a hacksaw to cut a piece of raceway to length for each segment (subtract 3 inches for the corner coupling at corner brackets).

• Push bushings into the raceway channels at both ends to protect wire insulation.

• Fit one end of the raceway channel onto the tongue of the extension plate at the existing box, and slip the other end into the corner bracket.

• Snap a mounting plate for the new outlet to one end of another length of raceway. Fit the other end to the corner bracket *(right).*

• With pliers, twist off any unused tongues from the mounting plate, and attach the plate to the wall *(inset).*

## 3. Fishing the New Wires

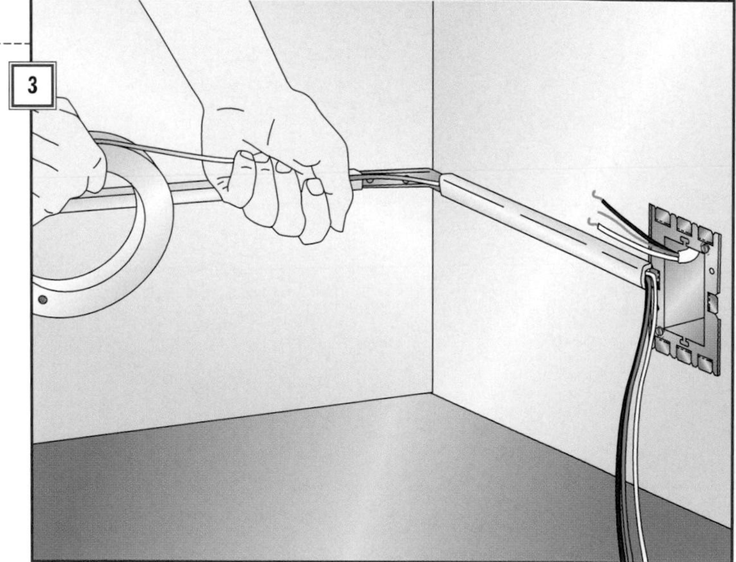


Individual wires, rather than cable, are used with surface-mount wiring systems.

• Starting at the corner bracket, feed a fish tape into the raceway channel and push it through to the existing box.

• Cut wires a couple of feet longer than the length of the entire run. Attach the ends to the fish-tape hook *(page 109).*

• Pull the tape and wires through the raceway to the corner bracket *(right),* and disconnect the wires from the fish tape.

• Feed the fish tape from the mounting plate at the new location to the corner bracket; attach the wires to it; and pull them through the raceway.

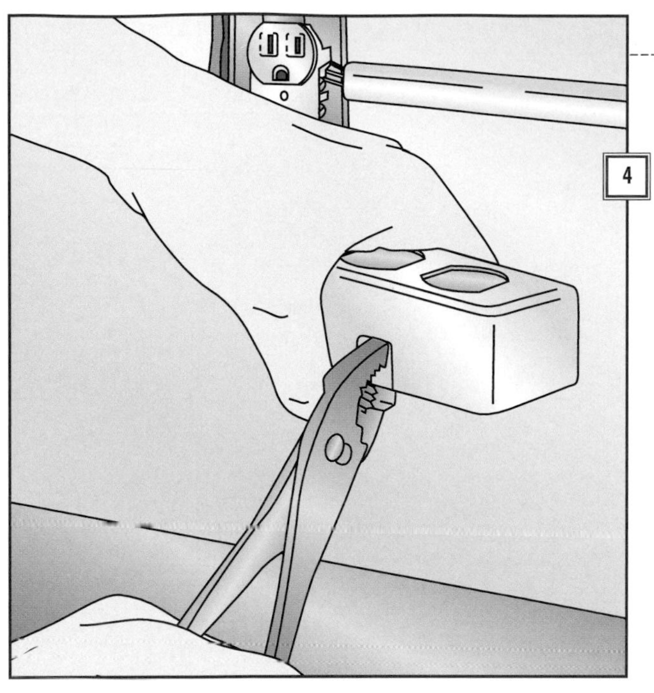


## 4. Installing the new outlet

- At the mounting plate for the new outlet, strip wire ends *(page 110)*. With long-nose pliers, form wire loops for terminal connections *(page 69)*. Then connect wires to the new outlet *(page 90)*.
- Screw the new outlet onto the mounting plate, making sure that it is straight.
- Identify the twist-out tab on the side of the outlet cover that corresponds to the position of the raceway. Remove the tab with pliers *(left)*.
- Snap the outlet cover onto the mounting plate and secure it with the mounting screw.

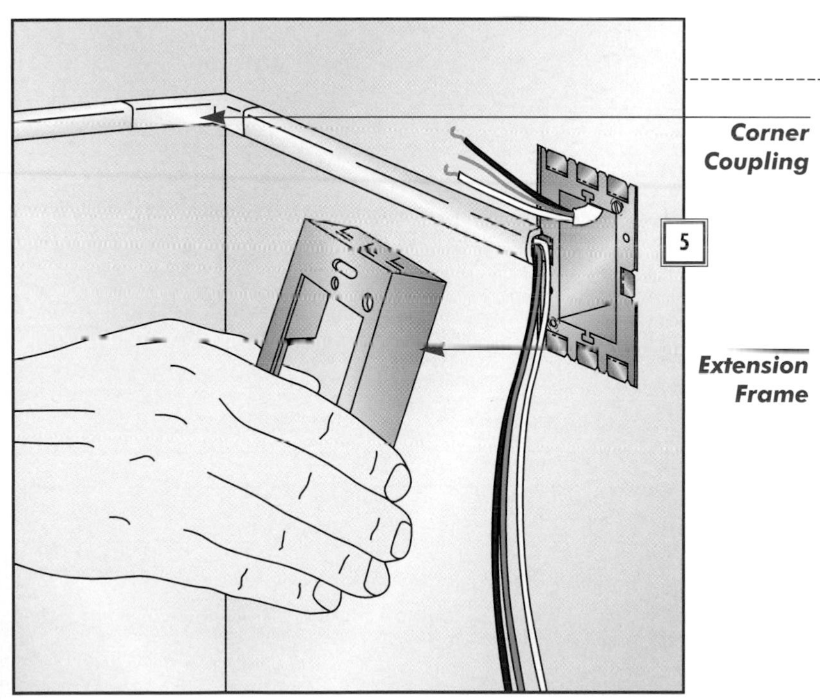


## 5. Completing the installation

- Snap the corner coupling onto the corner bracket, taking care not to pinch the wires.
- At the existing box, slip an extension frame onto the plate *(left)*, pulling the wires through the frame.
- Cut the ends of the new wires, leaving about 8 inches to make the necessary connections. Strip the ends and form loops for terminal connections *(page 69)*.
- Connect the old black wire to the top brass terminal of the outlet and the old white wire to the top silver terminal. Connect the new black wire to the bottom brass terminal and the new white wire to the bottom silver terminal. Prepare a ground jumper *(page 92)*, and twist it together with the ground wires to form a pigtail. Connect the ground jumper to the outlet's green ground terminal.
- Screw the outlet to the frame.
- Screw on the cover plate.

# Making the Right Connections

## Identifying the Incoming Hot Wire

When extending a circuit from a switch box, you need to know which black wire carries current from the service panel.

• With the power confirmed off *(page 68)*, separate the black wires and make sure they don't touch each other or the box. Turn on the power.

• Set a multitester to 250 in the AC-voltage range *(page 20)*. Touch one probe to a ground wire and the other to each black wire, in turn *(right)*. You'll get a reading when the probe touches the incoming wire. Turn off power and label that wire.

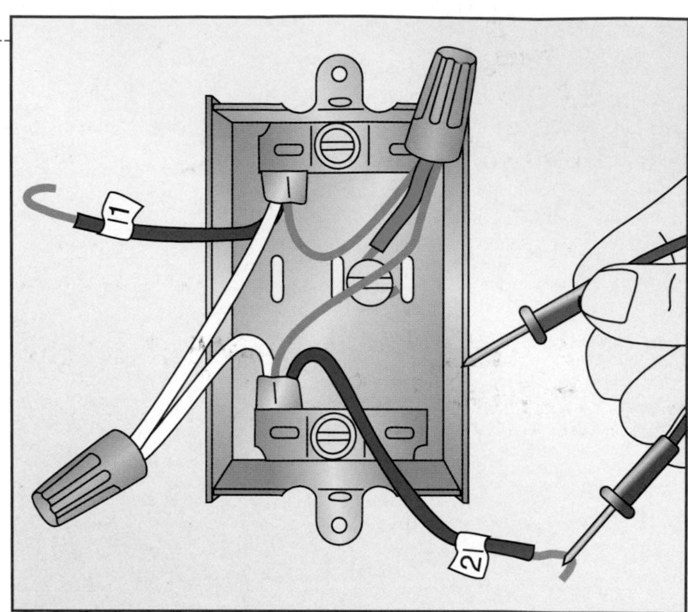

**CAUTION:** *This is a live voltage test. Be sure to hold the multitester probes by their insulated handles.*

## Extending from an End-of-Run Outlet

• Connect the black and white wires from each cable to brass and silver terminals directly opposite each other on the outlet. Join the new cable's ground wire to the other ground wires at the wire cap.

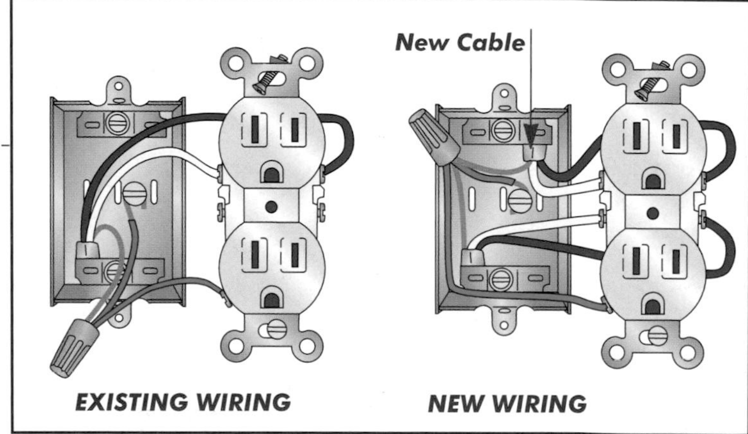


## Extending from a Middle-of-Run Outlet

• Connect the black and white wires from one cable to brass and silver terminals directly opposite each other on the outlet. Twist the other two black wires together with a jumper, screw on a wire cap, and attach the jumper to the remaining brass terminal. Twist the remaining white wires together with a jumper, and hook the jumper to the other silver terminal. Join the new cable's ground wire to the other ground wires at the wire cap.

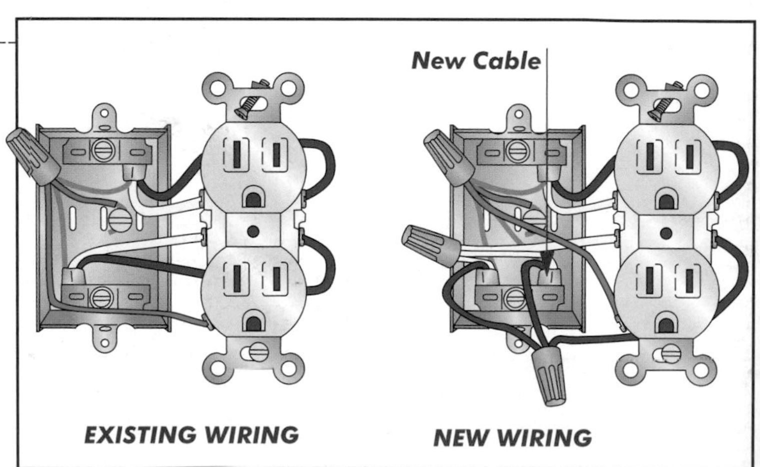

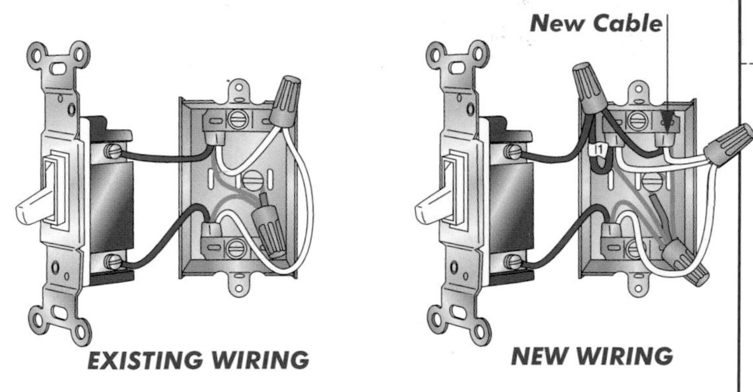


### Extending from a Middle-of-Run Switch

• Locate and tag the incoming hot wire *(page 120)*. Twist it together with the new hot wire and a jumper. Attach the jumper to a terminal. Attach the other black wire to the other terminal. Twist the white wire from the new cable together with other white wires. Twist the ground wire from the new cable together with wires in the ground pigtail *(left)*.

## Electrical Boxes

All wire connections—switches, receptacles, lights, junctions—must be made within code-approved boxes. Boxes come in plastic and galvanized steel, in several sizes and shapes *(below)*, and with a variety of mounting hardware.

How many wires you can pack into a box depends on its depth. "Number of Wires per Box," below, gives the maximum number of wires allowed in wall and ceiling boxes of various depths. Ground wires together count as one; jumper wires do not count. Any internal clamps, studs, or nipples count as one wire, as does an outlet or a switch.

**Side-Mounted Wall Box**

Mounts to a stud through screw holes at the side. Adjustable ears at the top and bottom hold box flush to wall.

**Mid-Wall Brackets**

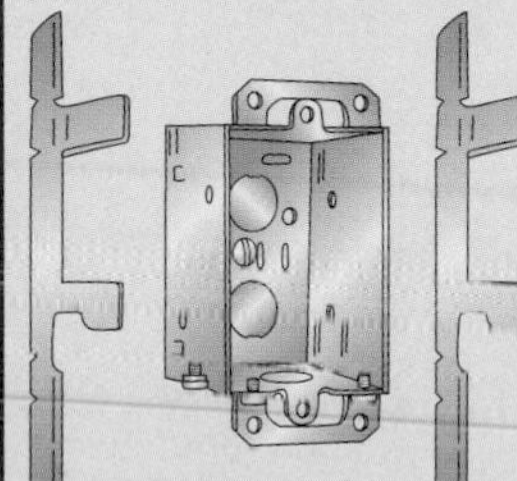

Use brackets to mount a wall box in an area of wallboard where there is no stud.

**Ganged Box**

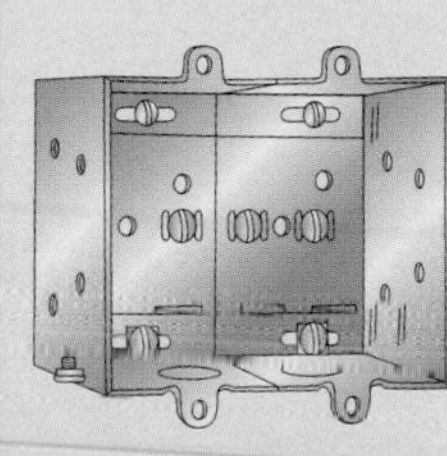

To increase capacity, remove right side of one box and left side of another; join with screws.

**Ceiling Box with Flange**

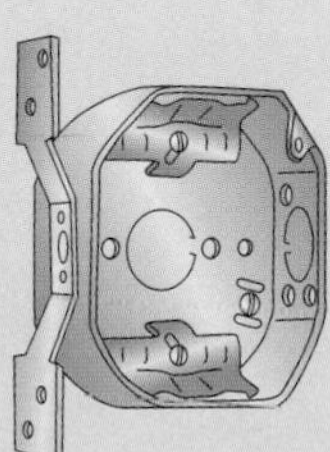

Screw the flange on the side of the box to the side of a joist, flush with the ceiling.

**Plastic Wall Box**

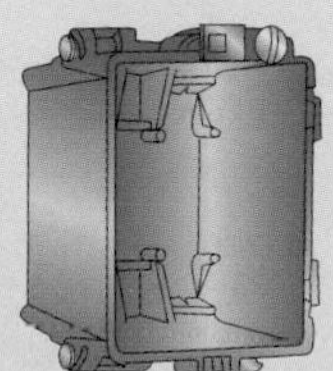

Mounts mid-wall or to studs with nails that fit into molded brackets at top and bottom. Check local codes to confirm that plastic boxes are permitted.

**Junction Box**

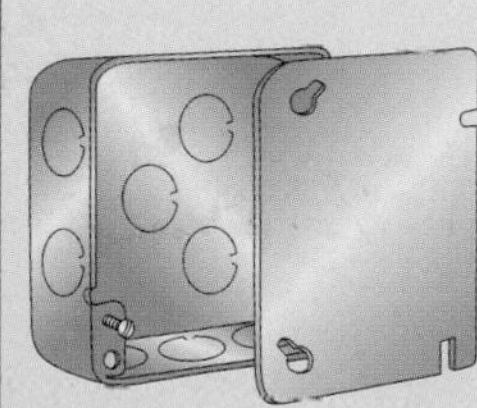

Use to enclose wiring splices; must be fastened in an accessible area.

**Number of Wires per Box***

| | Wall Box Depth | | | Ceiling Box Depth | | |
|---|---|---|---|---|---|---|
| | 2-1/2" | 2-3/4" | 3-1/2" | 1-1/4" | 1-1/2" | 2-1/8" |
| No. 10 | 5 | 5 | 7 | 5 | 6 | 8 |
| No. 12 | 5 | 6 | 8 | 5 | 6 | 9 |
| No. 14 | 6 | 7 | 9 | 6 | 7 | 10 |

**Consult local codes before extending household circuits.*

# FIX IT: Doorbells & Chimes

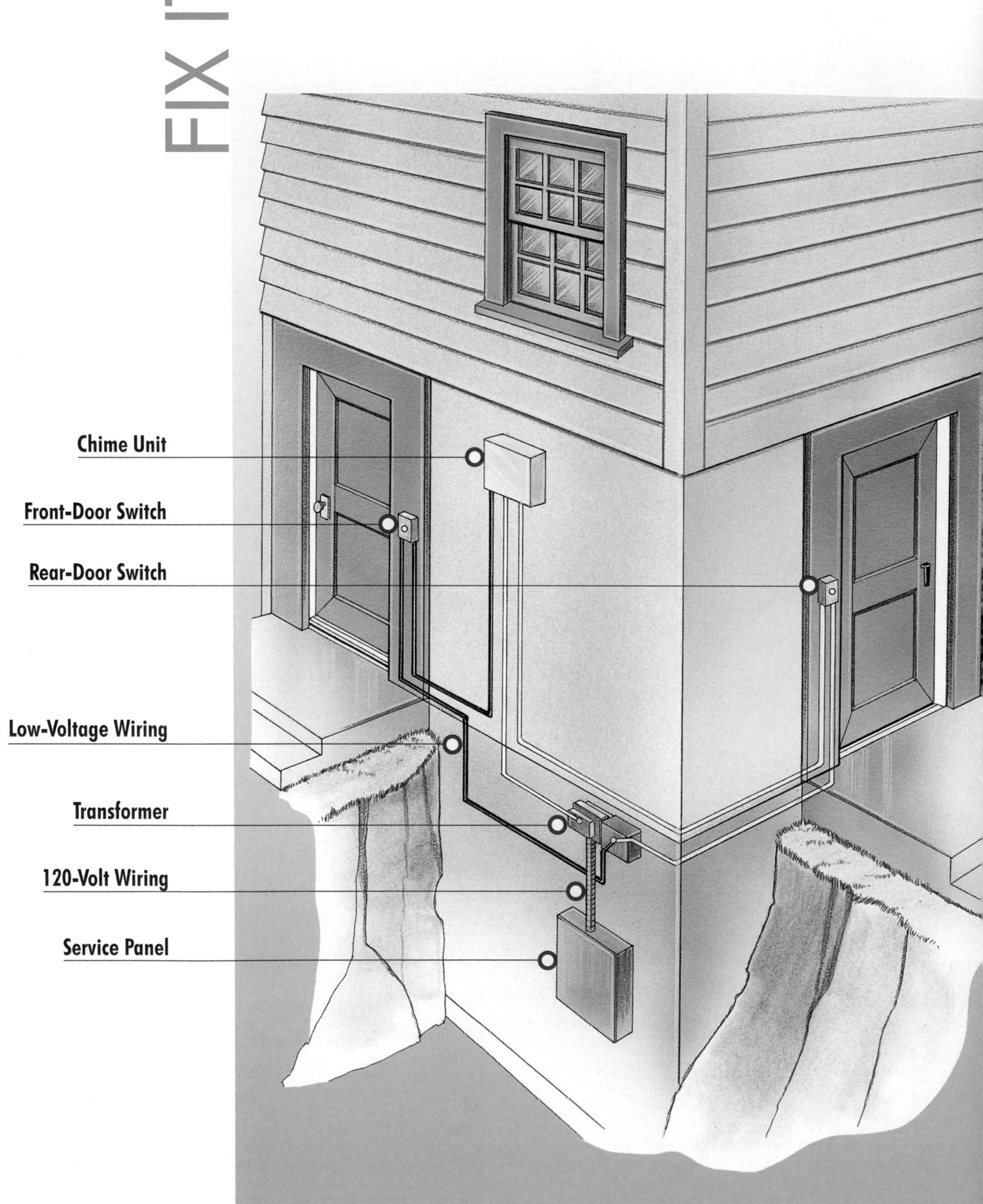

# Chapter 7

## Contents



## How They Work

**In a typical doorbell system, standard 120-volt current from the service panel is fed to a transformer, which reduces voltage to meet the requirements of the bell or chime unit. Bells usually work on 10 to 20 volts, chimes on 16 volts. Low-voltage current flows from the transformer to the bell or chime unit by way of push-button switches mounted at the doors. Pressing the button closes contacts within the switch, completing the circuit to sound the chimes or bell.**

# Troubleshooting

| Problem | Solution |
| --- | --- |
| • **Chime or bell unit does not ring** | Check for blown fuse or tripped circuit breaker **10** • Clean and adju push-button contacts **126** • Service switch-wire connections **126** • Replace faulty push button **127** • Service chime or bell connections **127** • Clean dirty chime or bell unit **127** • Replace faulty chime or bell **128** • Service transformer connections **128** • Test and replace faulty transformer **129** • |
| • **Chime or bell rings constantly** | Adjust push-button contacts **126** • |
| • **Chime fails to complete full cycle** | Clean chime unit **127** • Test and replace faulty transformer **129** • |

# Before You Start

When your doorbell isn't working, suspect a faulty push button. Subjected to frequent use and often somewhat exposed to the weather, push buttons tend to fail sooner than other doorbell-system components.

### A HIERARCHY OF PROBABLE CAUSES

After checking push-button contacts, inspect the chime or bell for a loose connection or faulty sounding components. Lubricating a balky mechanical chime or bell unit can often make it work again. Electronic units don't offer this opportunity. If your sounding unit is broken, it may be less expensive to replace it than to have it repaired.

If neither the push buttons nor the sounding unit are to blame for a silent doorbell, have a look at the transformer *(page 128)*. This electrical component has no moving parts, but they sometimes burn out. The wires that link the components together are the least likely cause of trouble. To check a wire, disconnect it at both ends and substitute a length of new wire along a convenient route—out a door and through a basement window to the transformer, for example. If the new wire solves the problem, find a less conspicuous route for it.

## Before You Start Tips:

- It's easier and faster to staple bell wire along baseboards and door casings than to fish it through walls.
- Replace a bell, chime, or transformer with a unit that matches the original in amperage and voltage ratings.

Screwdriver
Multitester
Wire strippers or multipurpose tool
Utility knife

Fine sandpaper
Cotton swabs
Rubbing alcohol
Masking tape

Chimes and bells normally operate on low voltage, so there is little shock danger. However, when working on a doorbell circuit, always turn off power at the service panel to protect yourself from the potential consequences of ground faults and miswiring, especially when you're working on connections at the transformer.

# Push-Button Repairs

## 1. Servicing the contacts

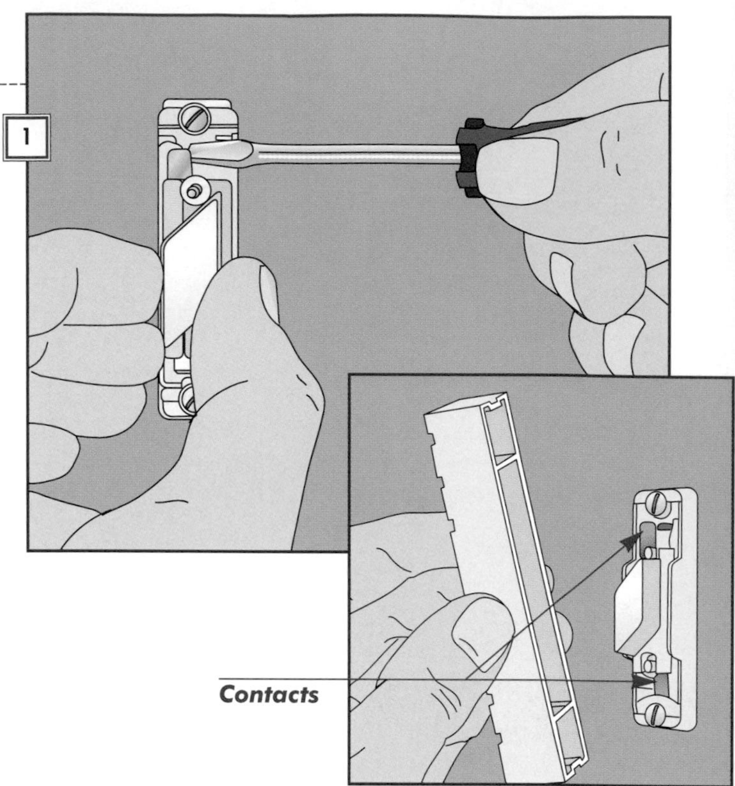


• Turn off power to the circuit by removing the fuse or tripping the circuit breaker *(page 10).*

• Pry off or unscrew the push-button cover and set it aside *(inset).*

• Use fine sandpaper to clean the metal contacts.

• Pry up the contacts with a screwdriver to improve the connection *(right).*

• Replace the cover, restore power to the circuit, and test.

• If the bell rings or the chime unit hums continuously, you've bent the contacts too far. Turn off the power to the circuit, remove the push-button cover, and bend the contacts down slightly. If the bell doesn't ring at all, go to Step 2.

## 2. Remaking the connections

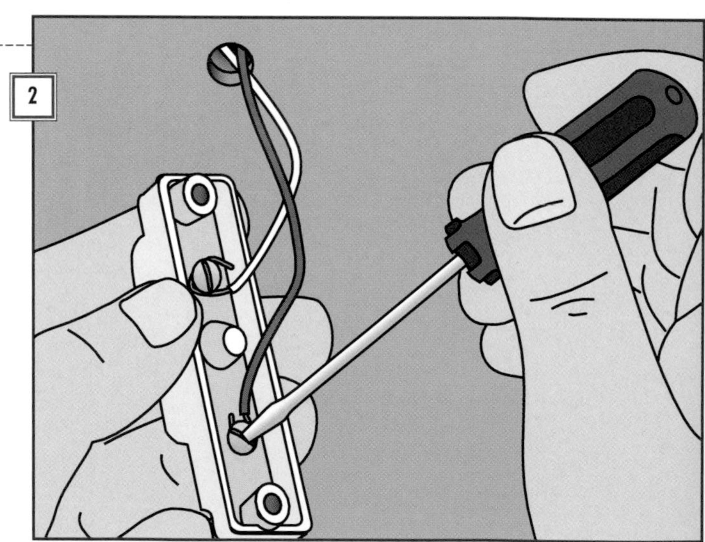


• Turn off power to the circuit at the service panel *(page 10).*

• Pry off or unscrew the push-button cover and set it aside.

• Loosen the screws securing the mounting plate to the door frame or siding. Pull the mounting plate forward to expose the wiring.

• Loosen the terminal screws and unhook the wires *(right).*

• Clip the exposed wire ends, strip back the insulation *(page 69),* and reattach the wires to the terminals.

• Reinstall the push button and restore power. If the bell doesn't sound when you push the button, proceed to Step 3.

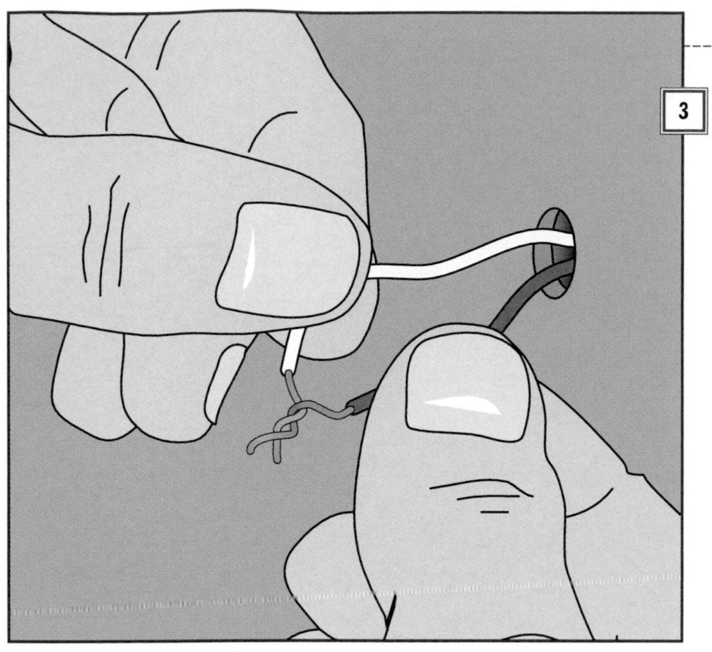


### 3. Isolating the Problem

• Turn off power to the circuit at the service panel; then loosen the terminal screws and unhook the wires as described on the opposite page.

• Twist the wires together *(left)* and restore power to the circuit. If the bell now sounds, the push button is faulty. If the bell doesn't sound, test the transformer *(page 129).*

• To replace a faulty push-button switch, turn off power to the circuit and connect the wires to the terminal screws on the new mounting plate.

• Screw the mounting plate to the siding or door frame, and snap on the cover. Restore power to the circuit, and test.

## Fixes for Chimes and Bells

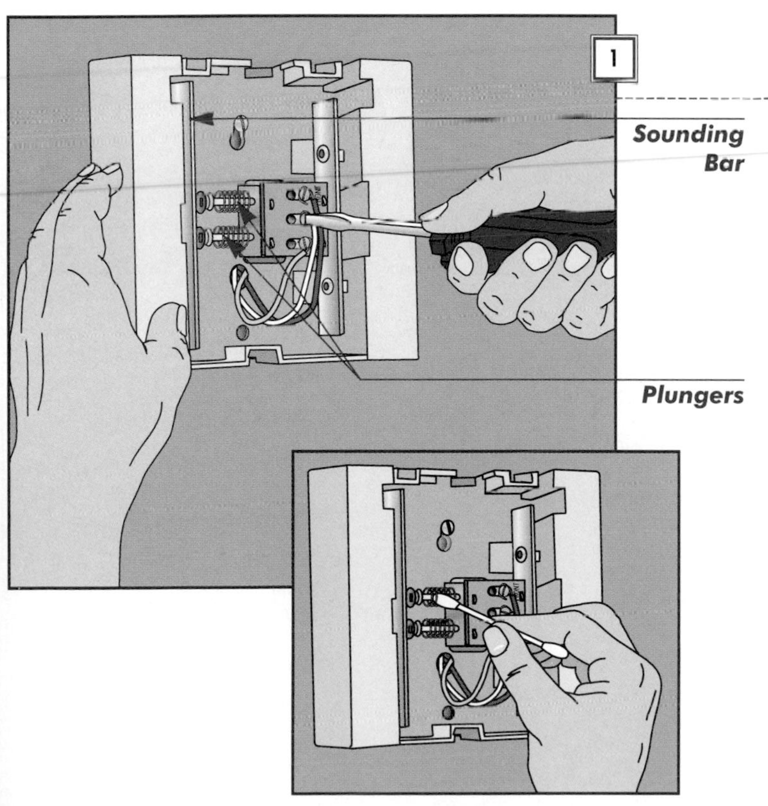


### 1. Servicing the Unit

The illustrations here and on the next page show a mechanical chime, but the sounding device may instead be a bell, a buzzer, or an electronic chime.

• Turn off power to the circuit at the service panel *(page 10).* Remove any screws holding the cover in place, and pull it forward to reveal the sounding components.

• Locate the terminals labeled FRONT, TRANS, and REAR. Clean and tighten loose connections *(left).* Cut back damaged wire ends, restrip them, and attach them to the terminals.

• On mechanical chimes, blow dirt away from the plungers and sounding bars. Clean the plungers with alcohol and a cotton swab *(inset).* Restore power and test. If there's no sound, go to Step 2.

### 2. Testing the Unit

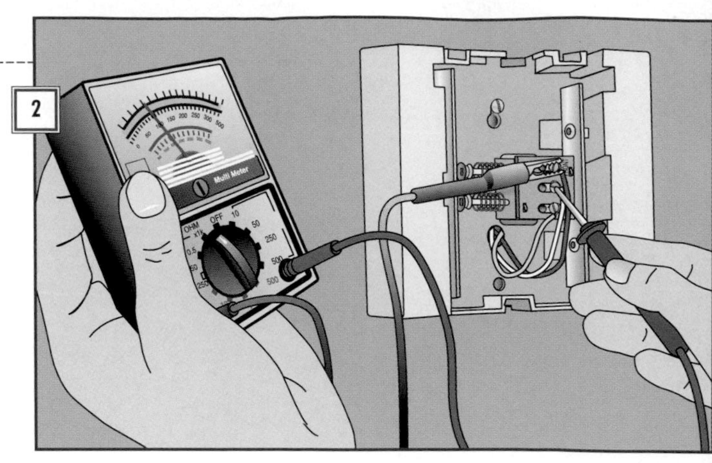


- Locate the transformer and check the low-voltage output on the specification label. Leave power to the circuit on, and set a multitester to 50 in the AC-voltage range.

- Holding the probes by their plastic covers, test a chime unit by clipping one probe on the TRANS terminal and touching the second probe to the FRONT terminal *(right).* Repeat the test with the REAR terminal, if present. (Test a bell unit by touching one probe to each bell terminal.)

- Readings within 2 volts of the transformer rating indicate sufficient voltage at the sounding unit; replace the unit if it doesn't ring. A reading of 0 volts indicates a break in the wiring; turn off power to the circuit, and replace faulty wires.

### 3. Replacing the Unit

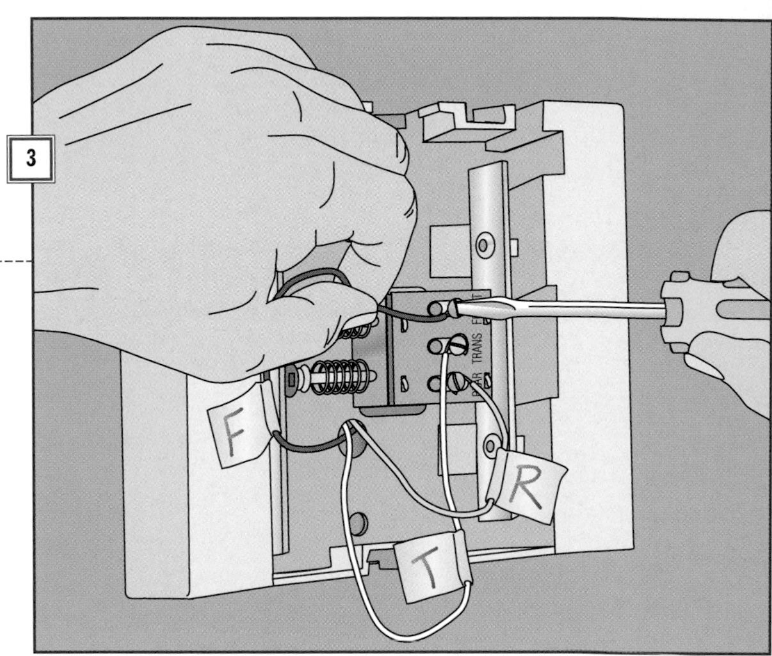


- Cut power to the circuit *(page 10).*

- Label the wires and disconnect them *(right).* Remove any mounting screws.

- Buy a replacement chime or bell that matches the transformer's voltage and amperage ratings. Mount it to the wall.

- Using labels on wires as a guide, attach the wires to the correct terminals of the new unit *(right).* Restore power and test.

## Replacing a Faulty Transformer

### 1. Checking the Low-Voltage Side

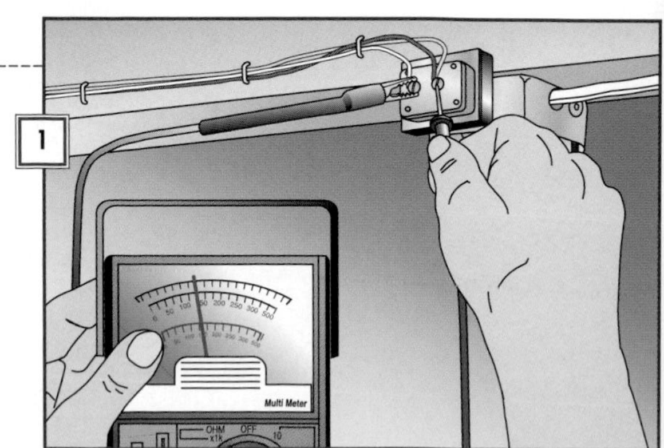


- Set a multitester to 50 in the AC-voltage range *(page 20).* Holding the probes by their plastic handles, touch one to each low-voltage terminal on the transformer *(right).* A reading that corresponds to the transformer rating indicates a faulty bell or chime. Go to Step 2 if there is no reading.

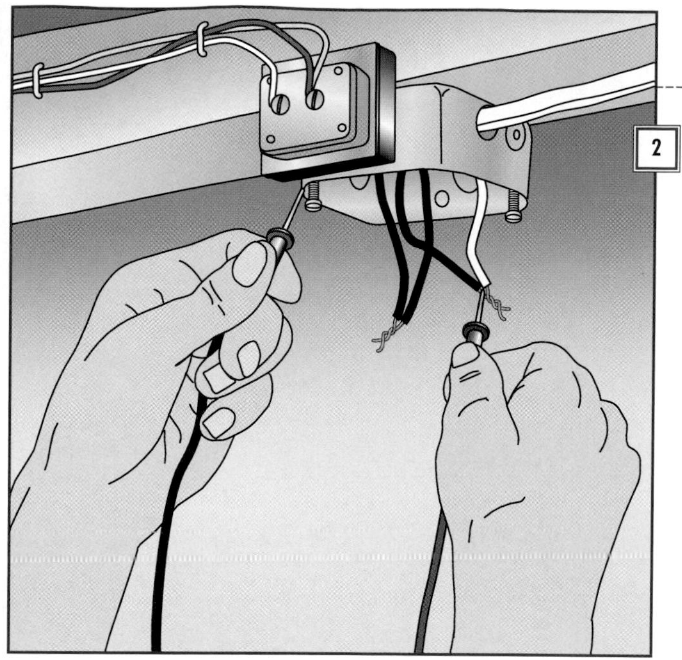

2

## 2. Servicing the 120-volt connections

• Cut power to the circuit at the service panel *(page 10)*. Unscrew the junction box cover to expose the 120-volt connections to the transformer.

• Unscrew the wire caps, taking care not to touch exposed wire ends. Set a multitester to 250 in the AC-voltage range. Touch one probe to the box if it's metal or to the ground screw if it's plastic; touch the other probe to each wire in turn *(left)*. Next test between the two wire connections. If there's a voltage reading in any test, go to the service panel and cut power to the correct circuit.

• Burnish wire ends or clip off damaged ones, strip back the insulation *(page 69)*, and reattach them. Restore power and test. Install a new transformer if the bell or chime does not ring *(Step 3)*.

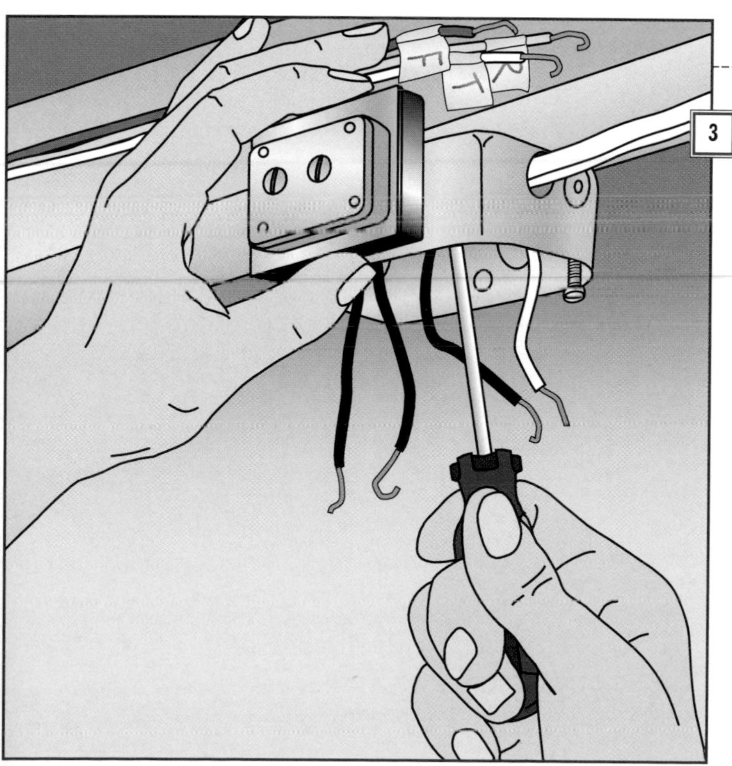

3

## 3. Replacing the transformer

• Cut power to the circuit at the service panel *(page 10)*. Test for voltage at the junction box *(Step 1)* to be sure the circuit is dead.

• Label the wires connected to the terminal screws on the low-voltage side of the transformer, and detach them.

• Disconnect the wires inside the junction box and unscrew the transformer *(left)*. Buy a replacement of the same voltage and amperage ratings.

• Thread the transformer leads through the side of the junction box, and fasten the transformer securely to the box.

• Twist one transformer lead to the black house wire and the other to the white house wire, securing each connection with a wire cap. Following their label instructions, attach the low-voltage wires to the transformer terminals, then restore power and test. Check the chime or bell unit if it does not ring *(page 127)*.

Chapter 8

# FIX IT: Outdoor Lighting

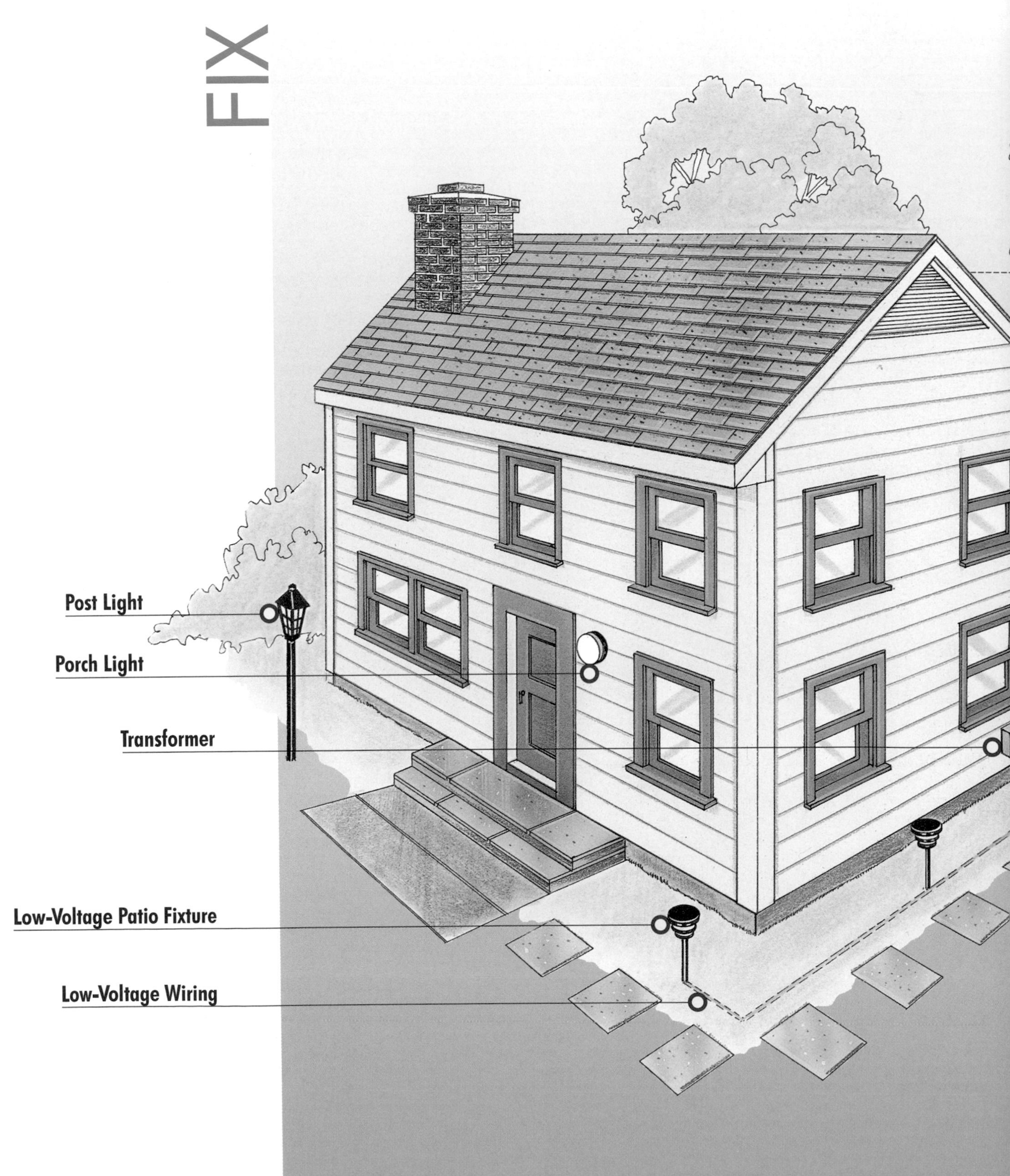

# Chapter 8

## Contents



## How It Works

Outdoor lighting comes in many forms. Some fixtures, such as the porch light attached to a home's exterior wall and the freestanding post light in the illustration at left, operate on 120-volt current fed directly from the service panel.

But many outdoor fixtures—the patio lights shown here, as well as path lights, spotlights, and accent lights that may be installed for safety and decoration—are available in low-voltage versions. Low-voltage systems receive power via a transformer that steps down 120-volt current to 12 volts.

In contrast to 120-volt outdoor circuits, which must be run be run with UF cable buried at least 6 inches deep (check your local electrical code), low-voltage wires may be covered with an inch or so of soil. Most low-voltage fixtures have built-in stakes that are simply stuck into the ground—features that make low-voltage systems inexpensive and easy to work with.

# Troubleshooting

| Problem | Solution |
|---|---|
| • Outdoor fixture flickers or does not light | Tighten or replace the bulb •<br>Replace the fuse or reset the circuit breaker **10** •<br>Test switch; replace if necessary **68–70** •<br>Check socket contact tab **134** • Check socket connections **135** •<br>Test and replace the socket **135** • |
| • One light in low-voltage system flickers or does not light | Tighten or replace the bulb •<br>Test and replace the socket **139–140** •<br>Replace the wiring **140** • |
| • More than one light in low-voltage system flickers or does not light | Replace the fuse or reset the circuit breaker **10** •<br>Clean and tighten the connections **89** •<br>Check the transformer fuses **141** •<br>Test and replace the transformer **141** • |

# Before You Start

Porch lights and post lights are electrically the same as other lights around the house. Low-voltage lights are exceptionally easy to maintain and repair.

### Weatherwise wiring

Most problems in outdoor lighting fixtures result from exposure to weather. Unless tightly sealed, the metal parts on outdoor fixtures are susceptible to corrosion caused by moisture. When servicing an outdoor fixture, check the housing and gasket for signs of damage. If any part is cracked, corroded, or broken, replace it before the wiring inside is affected. Protect joints between wall-mounted fixtures and siding or masonry with silicone sealant. And periodically remove dirt, insects, leaves, and other debris that tends to collect inside fixtures in exposed locations.

## Before You Start Tips:

- Install weatherproof bulbs in outdoor fixtures. They resist the effects of rain, snow, and ice and tend to need replacement much less frequently than standard bulbs.
- Apply a thin coat of petroleum jelly to the threads of outdoor bulbs before screwing them into their sockets: They will be much easier to remove when they burn out.
- Wrap wire-cap connections with plastic electrical tape to protect the wires from moisture.
- Protect 120-volt outdoor circuits with a ground-fault circuit interrupter (GFCI).

Wire strippers
Multitester
Screwdrivers
Long-nose pliers
Lineman's pliers

Wire caps
Electrical tape
Sandpaper
Silicone sealant

Turn off power to the circuit, even if you're only cleaning or changing a light bulb. Stand on a wooden plank or rubber mat when working on damp ground. Use a wooden or fiberglass ladder to get at hard-to-reach fixtures.

Before servicing a low-voltage system, unplug the transformer from the 120-volt household current.

# Porch Lights

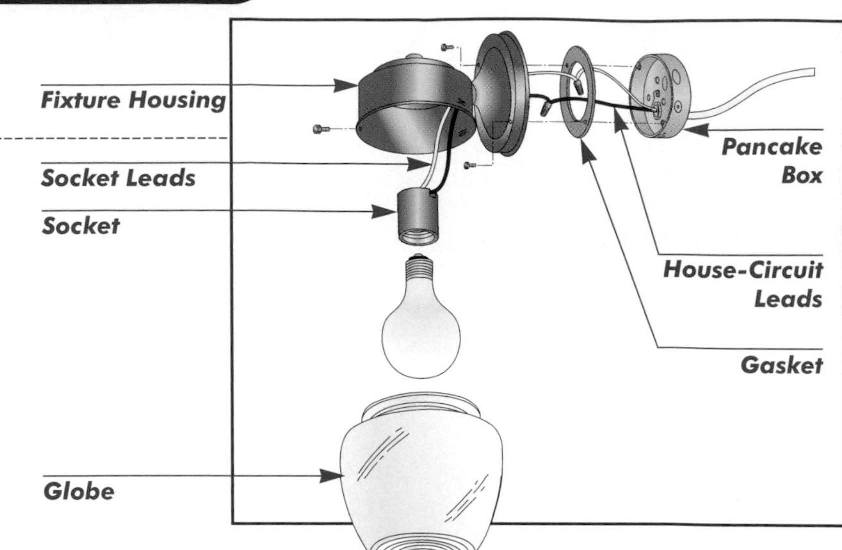


## Porch-light anatomy

A porch light may be mounted to a standard wall box or to a pancake box *(right)*. While the components and wiring are the same as for an interior 120-volt fixture, a gasket must be installed between the fixture and the box to protect the wiring from moisture.

## 1. Testing for voltage

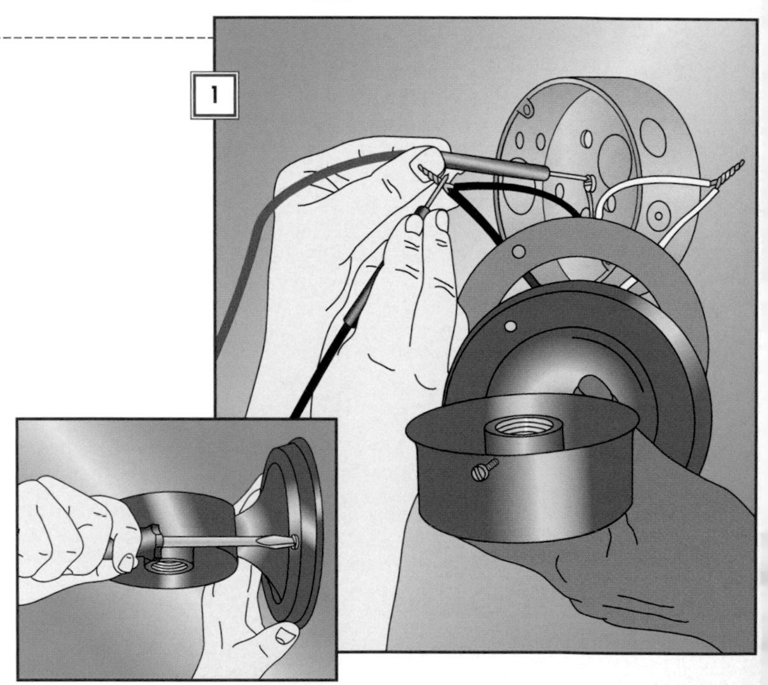


- Cut power to the circuit at the service panel *(page 10)*.
- Remove the globe and bulb, and unscrew the fixture from the wall *(inset)*. Have a helper hold the fixture while you pull the wires out of the box.
- Unscrew the wire caps, taking care not to touch the exposed connections.
- To confirm that power is off, set a multitester to 250 in the AC-voltage range *(page 20)*. Touch one probe to the black wires, and the other first to the ground terminal *(right)* and then to the white wires. Next test between the white wires and the box.
- If any test shows voltage, cut power to the correct circuit at the service panel. If there is no voltage, untwist the connections and take down the fixture.

## 2. Refurbishing the socket

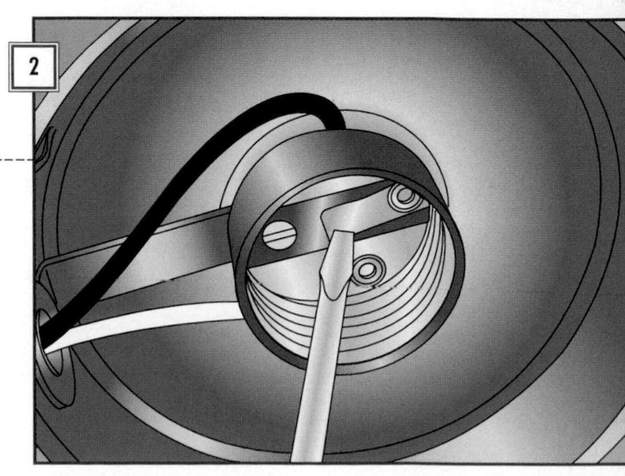


- Inspect the socket for wear or corrosion.
- Scrape the contact tab clean with a screwdriver or an old knife *(right)*.
- Bend the tab up slightly to improve contact with the bulb *(page 18)*.

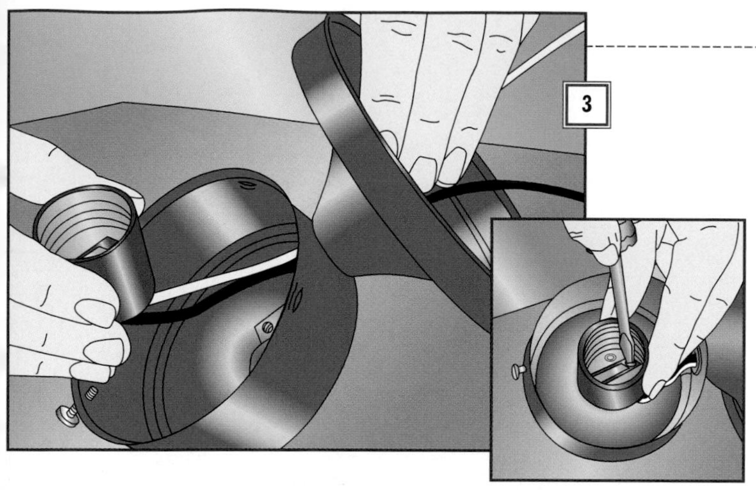


### 3. Checking socket connections

- Remove the socket-mounting screw *(inset).* Pull the socket from the fixture to expose the wire connections *(left).*

- If the socket has terminal screws, loosen them and detach the leads before proceeding to Step 4.

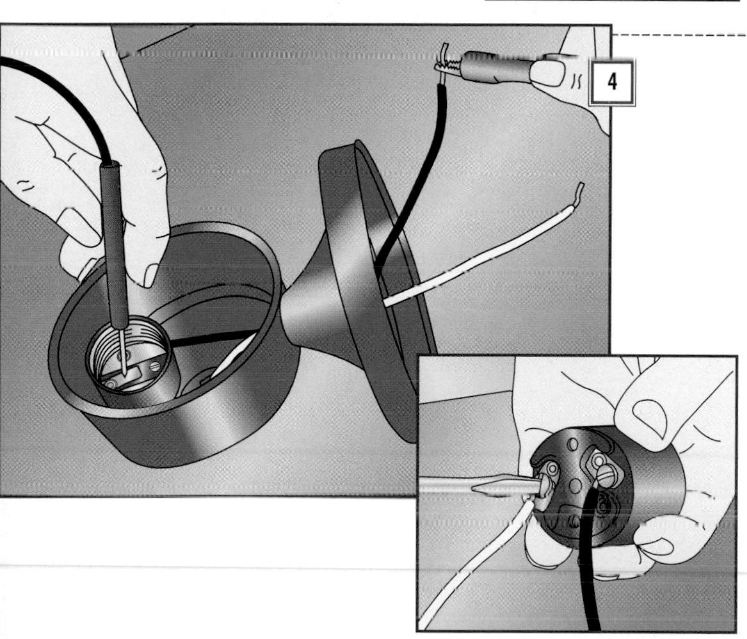


### 4. Testing and replacing the socket

- Set a multitester to RX1 *(page 20).* Test between the black wire (or brass terminal screw) and the socket's contact tab *(left).* Then place the probes on the white wire (or silver screw) and the socket's threaded tube. Both tests should show continuity.

- If so, reinstall the socket, cleaning terminals and exposing fresh wire ends as needed *(page 89).* For a socket with screw terminals, connect the black wire to the brass terminal, and the white to the silver one *(inset).*

- If the socket fails the test, replace it with a new one that fits the fixture. Feed the wires through the fixture, and tighten the socket-mounting screw.

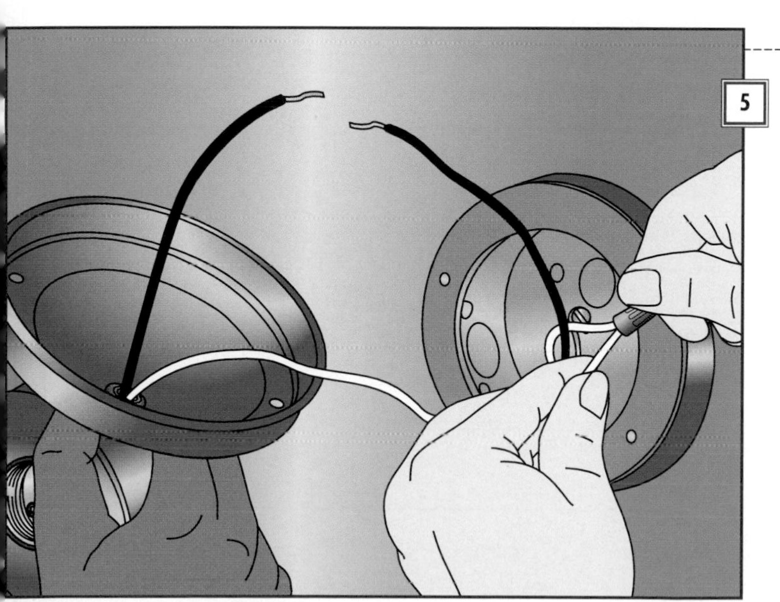


### 5. Remounting the lamp

- Trim and restrip the wires in the box *(page 110).* If the gasket is worn, replace it.

- While a helper holds the fixture, use wire caps to connect the white socket lead to white house wire and the black lead to the black house wire *(left).* Wrap the caps with electrical tape to protect against moisture, then gently fold the wires into the box.

- Set the fixture against the wall and tighten the mounting screws. Reinstall the bulb and globe, then turn on the power.

- With a caulking gun, lay a thin bead of silicone sealant around the fixture base.

# Post Lamps

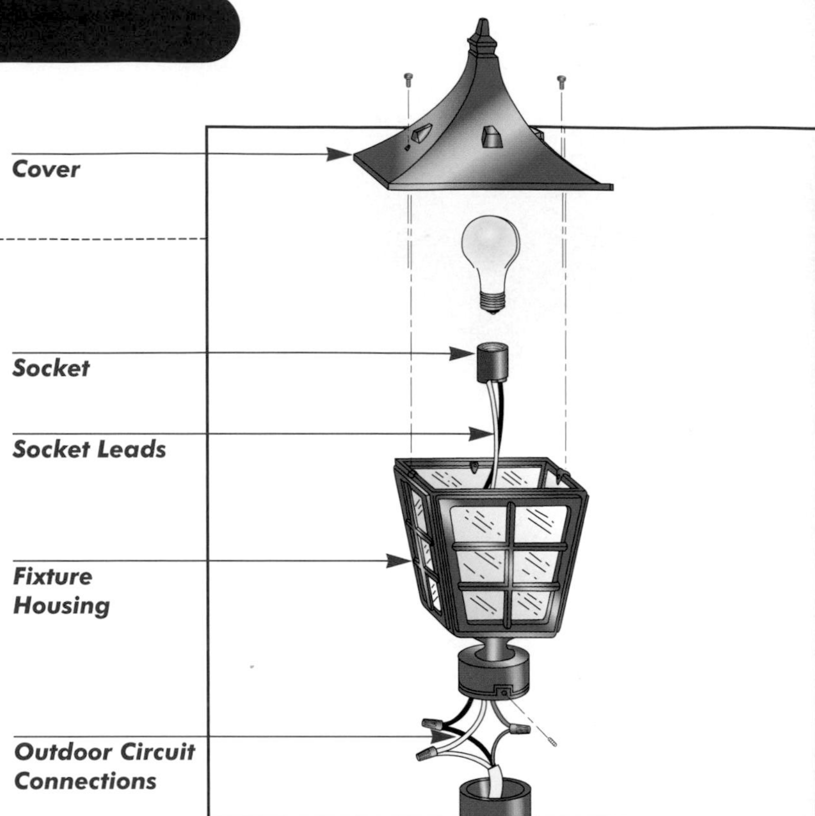


## POST-LAMP ANATOMY

A typical post light *(right)* consists of a socket and bulb inside a weathertight housing. The assembly is mounted atop a hollow post anchored in concrete. The wiring for a 120-volt lamp runs through the post and buried metal conduit to the service panel.

## 1. GAINING ACCESS AND TESTING FOR VOLTAGE

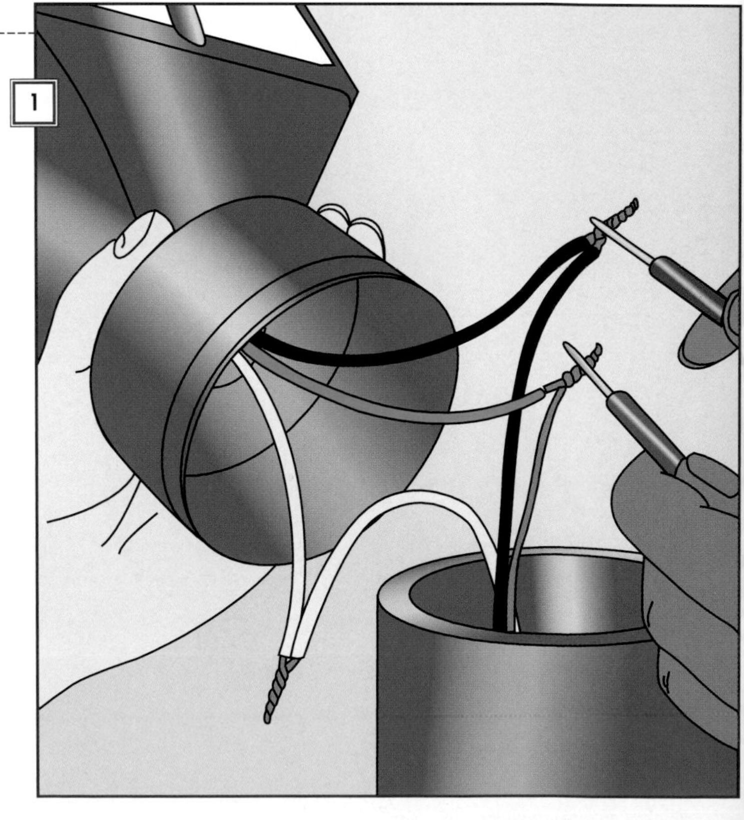


- Cut power to the circuit at the service panel *(page 10).*
- Take out the cover screws and set the cover aside. Remove the bulb, as well as any glass or plastic panes in the housing.
- Loosen the screws securing the fixture to the post. Lift the fixture, exposing the wire connections. Have a helper hold the fixture while you unscrew the wire caps; take care not to touch exposed wire ends.
- With the multitester set to 250 on the AC-voltage scale *(page 20),* place one probe on the black wire connection and the other on the ground connection *(right).* Then touch probes to the black wire connection and the white wire connection. Next test between the white wires and the ground-wire connection.
- If you get a voltage reading in any of the tests, return to the service panel and shut off the correct circuit.

## 2. Disconnecting the Housing

• When the power is off, separate the wires and tape the incoming wires over the edge of the post to prevent them from falling into the post *(left).*

• Set the fixture housing aside.

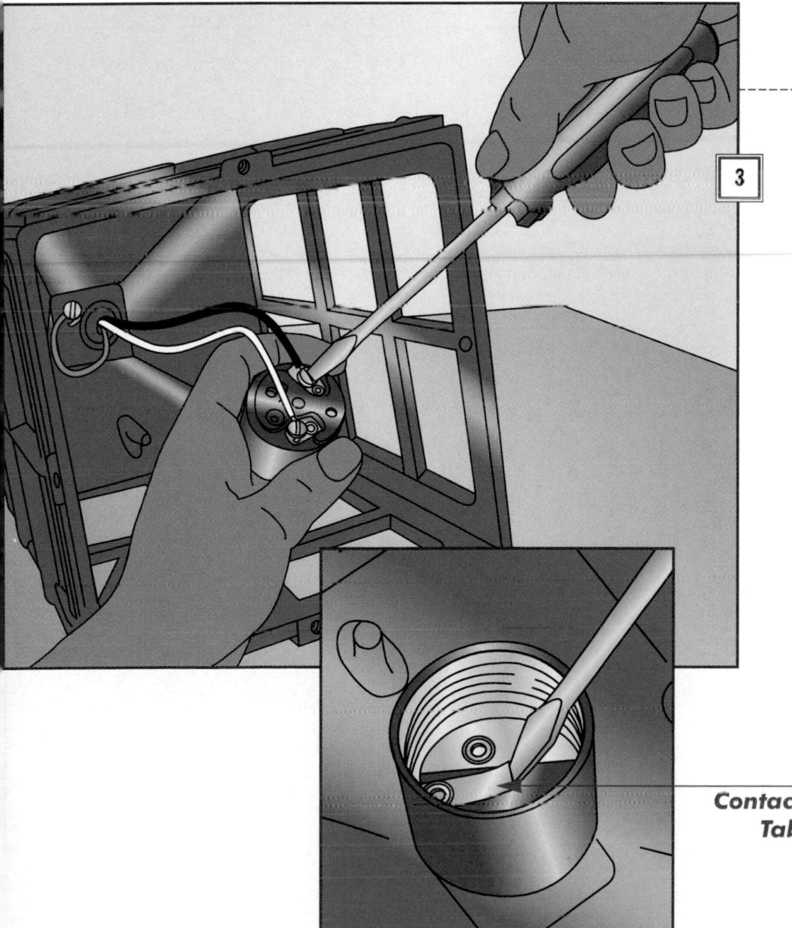


## 3. Cleaning and Adjusting the Socket

• With an old knife or screwdriver, scrape any corrosion off the socket contact tab, then pry it up slightly *(inset)* to improve contact with the bulb.

• Loosen the socket-mounting screw, and pull the socket far enough out of the fixture to expose the lead connections *(left).* If the leads are connected to terminal screws, loosen them and detach the leads before proceeding to Step 4.

## 4. Testing the socket

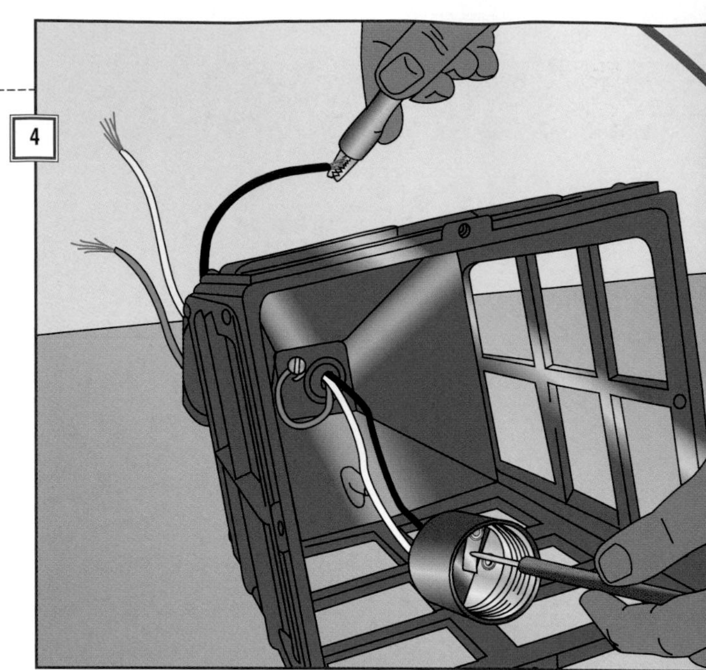


• Set a multitester to RX1 *(page 20).* Test between the black wire (or brass terminal screw) and the socket contact tab *(right).* Then place the probes on the white wire (or silver screw) and the socket's threaded tube. Both tests should show continuity. If not, go to Step 5.

• Reinstall the socket if it passes the test, cleaning terminals, cutting off old wire ends, and stripping new ones as needed *(page 89).* For a socket with terminal screws, connect the black wire to the brass screw and the white wire to the silver one.

## 5. Replacing the socket

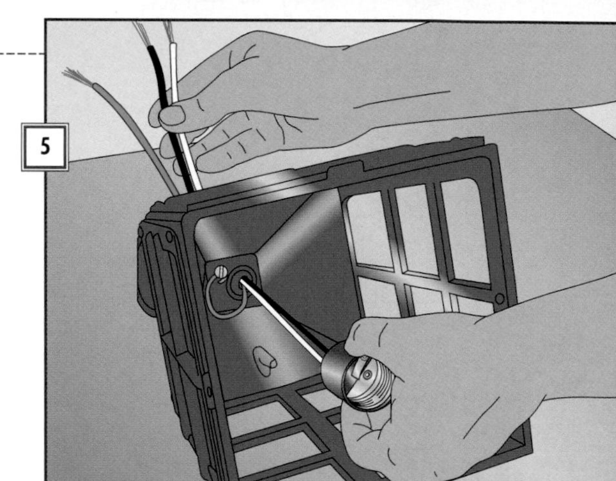


• Buy a new socket that fits the fixture housing. If it has terminal screws, attach the leads as described in Step 4.

• Install the socket by feeding the wires into the fixture housing and pulling them through the bottom *(right).*

• Tighten the socket-mounting screw.

## 6. Mounting the post light

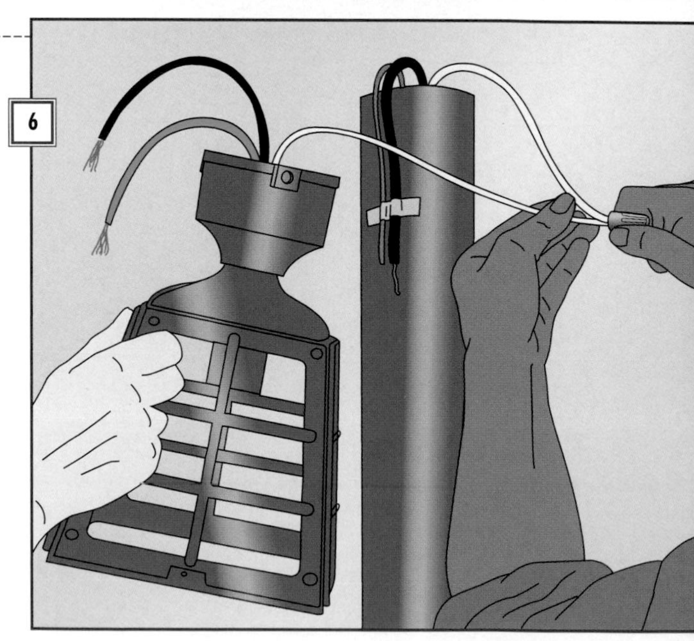


• Trim and restrip the wires inside the post to improve the connections *(page 110).*

• While a helper holds the fixture, twist together the ends of the white wires and screw on a wire cap *(right).* Repeat for the black wires, then for the grounding wires.

• Wrap the wire caps with electrical tape to protect them from moisture, and gently fold the wires into the post.

• Position the fixture on the post, and tighten the retaining screws. Then screw in a bulb, restore power, and test the fixture.

• Slide in the glass panels, if any, and reinstall the cover.

# Low-Voltage Garden Lights

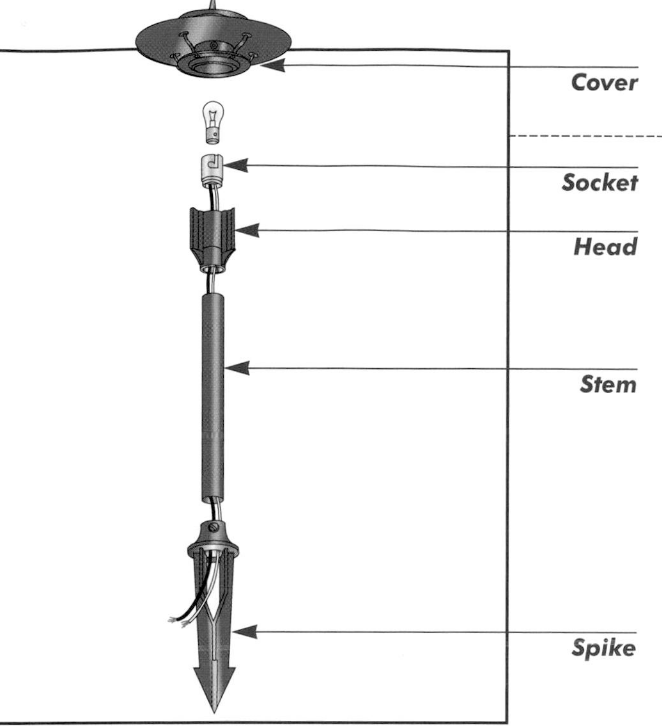


## Low-voltage light anatomy

A typical low-voltage path light *(left)* includes a cover, a bayonet-style socket designed for push-in bulbs, a hollow stem, and a spike that's driven into the ground. Wires run from the socket down through the stem, then to the next light in the series or to a transformer *(page 130).*

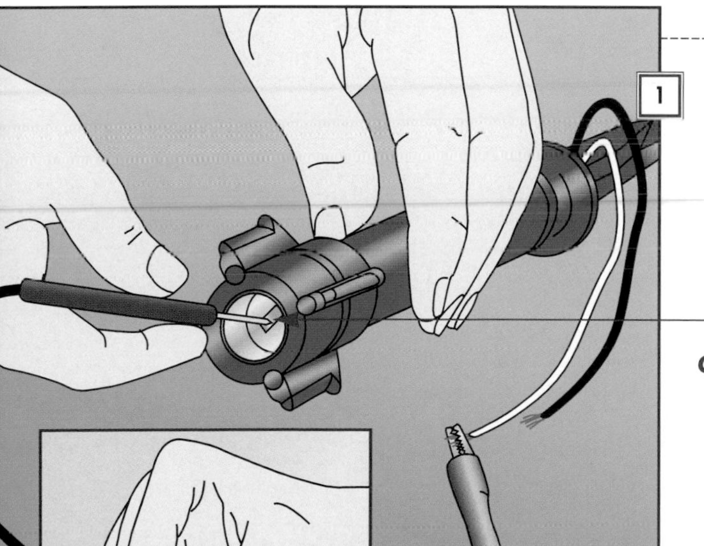


## 1. Replacing the bulb and testing the socket

- Unplug the transformer.
- Unscrew and remove the light's cover.
- To replace a burned-out push-in bulb, gently depress the bulb and turn it counter-clockwise, then lift it out *(inset).* Install a replacement. If the new bulb doesn't light, test the socket as follows.
- Pull the light out of the ground, then withdraw the wires from the base of the stem to expose the connections.
- Unscrew the wire caps and detach the wires.
- Set a multitester to RX1 *(page 20).* Clip one lead to the white wire and touch the other to the metal socket tube. Then test between the black wire end and the socket contact tab *(left).* If the meter doesn't indicate continuity in both cases, buy a compatible replacement socket with pre-attached wires.

## 2. Installing a new socket

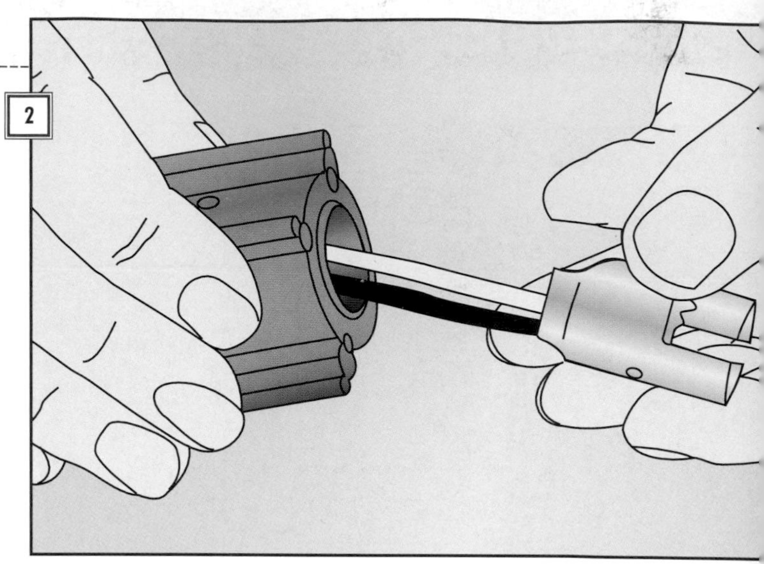


- Grasp the rim of the old socket with long-nose pliers, and pull to loosen it; then extract it from the stem by hand *(right)*. If the socket is stuck, separate the head of the light from the stem, then force the socket up from below with a screwdriver.

- Feed the wires of the new socket through the stem, then push the socket home.

- Check for dirty or corroded wire ends. Clean or clip them off as needed *(page 89)*.

- Reconnect the wires, reassemble the light, and plant it in the ground.

- Turn on the power. If the bulb doesn't glow, inspect the wires between the lights for damaged insulation, breaks, and loose connections. Exposed wiring is subject to moisture damage.

## 3. Replacing low-voltage wires

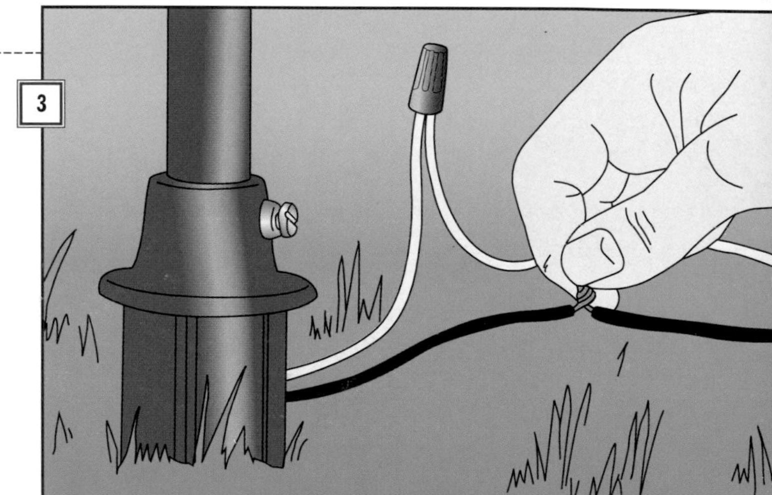


- To replace a wire, unplug the transformer and lift the fixtures joined by the defective wire a few inches out of the ground. Then pull the wires out of the base of the stem, and undo the connections.

- Run replacement wire of the same gauge along the surface or, for better protection, bury it 6 inches under the soil surface.

- Strip back the wire ends *(page 110)*. Twist each wire end together with a matching-color wire from the light stem *(right)*, and secure the connections with wire caps.

- Plug in the transformer and test the light.

# Troubleshooting a Transformer

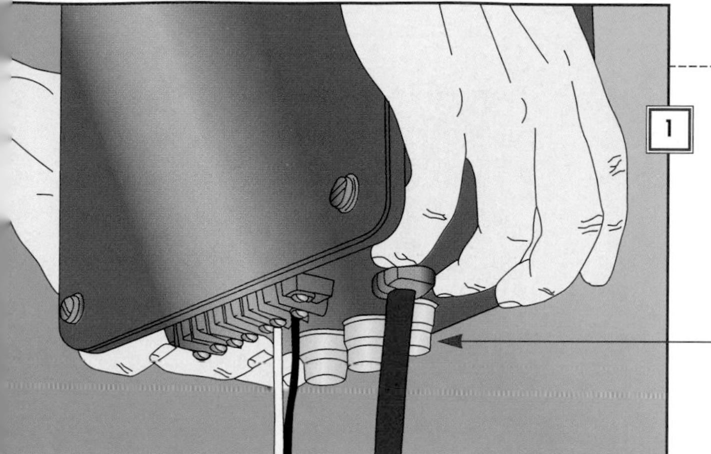


### 1. Dismounting the Transformer

- Unplug the transformer from the outlet.
- Lift the case off its bracket *(left).* Set it down on a flat, dry surface with the fuses facing upward.
- Inspect the cord and plug. If they look damaged, replace the transformer.

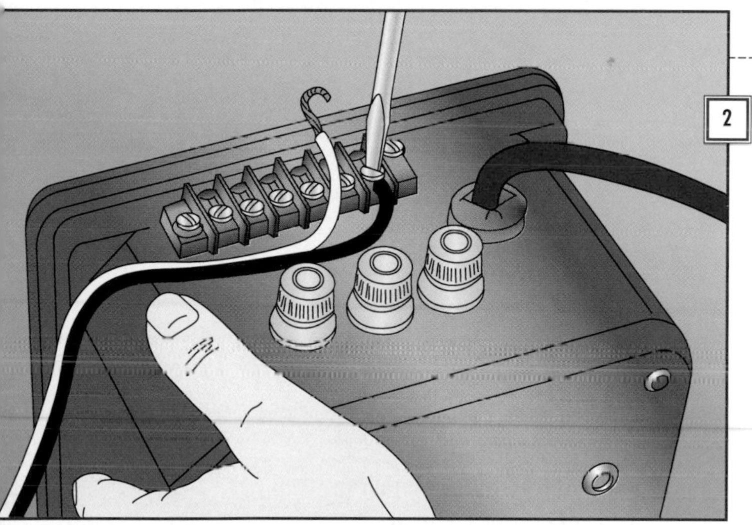


### 2. Improving Connections

- Disconnect the wires from their terminals on the transformer *(left).*
- If the wires or terminals are dirty or corroded, clean them with fine sandpaper *(page 89),* or strip the insulation to make a fresh connection *(page 110).*
- Reconnect the wires to the terminals.
- Plug in the transformer, and flip on the switch that controls the lights. If they don't light, test the transformer *(Step 3).*

### 3. Testing and Replacing the Transformer

- Remove the fuse by pressing it in slightly and turning it counterclockwise.
- If it looks blown or damaged, replace it. If not, reinstall the fuse, then plug the transformer into the outlet.
- Set a multitester to 50 in the AC-voltage range *(page 20).* Holding the probes by their insulated handles, place them on the terminal screws *(left).*
- If the voltage reading differs from the transformer rating by more than 10 percent, buy a new transformer.

# Index

0874

Time-Life Books
is a division of Time Life Inc.

## Time Life Inc.

**George Artandi**
President and CEO

## Time-Life Books

**Neil Kagan**
Publisher/Managing Editor

**Steven A. Schwartz**
Vice President, Marketing

## How To Fix It:

### Lighting & Electricity

**Lee Hassig**
Editor

**Wells P. Spence**
Director of Marketing

**Kate McConnell**
Design Director

**Janet Johnson**
Special Contributor (design)

**Christopher Hearing**
Director of Finance

**Marjann Caldwell**
**Patricia Pascale**
Directors of Book Production

**Betsi McGrath**
Director of Publishing Technology

**John Conrad Weiser**
Director of Photography
and Research

**Barbara Levitt**
Director of Editorial Administration

**Gertraude Schaefer**
Production Manager

**James King**
Quality Assurance Manager

**Louise D. Forstall**
Chief Librarian

## Butterick Media

### Staff for Lighting & Electricity

**Michael Chotiner**
Editor

**Daniel Newberry**
Associate Editor

**Charles Wardell**
Writer

**Caroline Politi**
Director of Book Production

**Robyn Shockley**
Managing Editor

**Ben Ostasiewski**
Art Director

**David Joinnides**
Page Layout

**Jim Kingsepp**
Technical Consultant

**Barbara M. Webb**
Copy Editor

**Nan Badgett**
Indexer

**Lillian Esposito**
Production Editor

**Art Joinnides**
President

## Picture Credits

**Brian Kraus**
**Juan Rios**
**"Butterick Media"**
Photography

**Mario Camacho**
**Bob Crimi**
**Geoff McCormack**
**Linda Richards**
**Joseph Taylor**
Illustration



First printing. Printed in U.S.A.
Published simultaneously in Canada.
School and library distribution
by Time-Life Education,
P.O. Box 85026, Richmond, Virginia 23285-5026.



**Library of Congress**
**Cataloging-in-Publication Data**
Lighting & Electricity/ by the editors of Time-Life Books
p. cm. -- (How to Fix It)
ISBN 0-7835-5653-5
**1. Electric lighting---Wiring--Amateurs' manual.**
**2. Electric light fixtures—Maintenance and repair--Amateurs' manual.**
**3. Electric wiring--Amateurs' manual.**
**I. Hazelton, Ron. II. Time-Life Books. III. Series.**
**TK9921.L55 1998 98-39615**
**621.319'24--dc21**
**CIP**